Supporting open minds since 2005

Search Algorithm - Essence of Optimization
http://dx.doi.org/10.5772/intechopen.87787
Edited by Dinesh G. Harkut

Contributors
Ahmed Tchvagha Zeine, Rachid Ellaia, Hamed Ould Sidi, Emmanuel Pagnacco, Santiago-Omar Caballero-Morales, Gladys Bonilla-Enriquez, Arshad Jamal, Hassan M. Al-Ahmadi, Meshal Almoshaogeh, Sajid Ali, Mudassir Iqbal, Farhan Muhammad Butt, Lahcen Amhaimar, Ali El Yaakoubi, Mohamed Bayjja, Kamal Attari, Saida Ahyoud, Dinesh G. Harkut, Prachi Thakar, Lovely Mutneja

Notice
Statements and opinions expressed in the chapters are these of the individual contributors and not necessarily those of the editors or publisher. No responsibility is accepted for the accuracy of information contained in the published chapters. The publisher assumes no responsibility for any damage or injury to persons or property arising out of the use of any materials, instructions, methods or ideas contained in the book.

First published in London, United Kingdom, 2023 by IntechOpen
IntechOpen is the global imprint of INTECHOPEN LIMITED, registered in England and Wales, registration number: 11086078, 5 Princes Gate Court, London, SW7 2QJ, United Kingdom
Printed in Croatia

British Library Cataloguing-in-Publication Data
A catalogue record for this book is available from the British Library

Additional hard and PDF copies can be obtained from orders@intechopen.com

Search Algorithm - Essence of Optimization
Edited by Dinesh G. Harkut
p. cm.
Print ISBN 978-1-83969-086-0
Online ISBN 978-1-83969-087-7
eBook (PDF) ISBN 978-1-83969-088-4

We are IntechOpen,
the world's leading publisher of Open Access books
Built by scientists, for scientists

6,200+
Open access books available

169,000+
International authors and editors

185M+
Downloads

156
Countries delivered to

Our authors are among the
Top 1%
most cited scientists

12.2%
Contributors from top 500 universities

Interested in publishing with us?
Contact book.department@intechopen.com

Numbers displayed above are based on latest data collected.
For more information visit www.intechopen.com

Meet the editor

Dr. Dinesh G. Harkut is a Head and Associate Professor in the Computer Science & Engineering Department, Prof Ram Meghe College of Engineering & Management (PRMCEAM), Badnera, India. A pioneering researcher in soft computing and embedded systems, he holds dual doctorates in Computer Science & Engineering and Business Management and has two registered patents to his name. His primary research interests are in computer artificial intelligence, big data, analytics, embedded systems, and e-commerce. He has supervised around eighteen master's degree and twenty-four bachelor's degree students. He has published forty-eight papers in refereed journals and published six books with international publishers. He is the principal investigator in establishing a center of excellence for renowned technological giants like IBM, Oracle, Texas Instruments, and Huawei at PRMCEAM. He is instrumental in setting up industry-funded laboratories at ARM, Cypress Semiconductor Corporation, Intel FPGA, Wind River, and Xilinx. Dr. Harkut is a fellow of the Institute of Electronics and Telecommunication Engineers (IETE), a life member of the Indian Society for Technical Education (ISTE), a senior member of the Universal Association of Computer and Electronics Engineers (UACEE), and a professional member of the International Association of Engineers (IAENG).

Contents

Preface

The advent and widespread application of computational intelligence in recent years, specifically artificial intelligence, machine learning, deep learning, data mining, and data analytics, has led to increased interest in sophisticated and optimized searching and data-matching problems, especially in information retrieval and computational biology.

This book explores how we search for resources in our minds and in the world and discusses the recent advancement and applications of quantum search, harmony search, cognitive search, genetic search, combinatorial search, and parallel random search algorithms in solving tough optimization difficulties that arise in advanced cross-domain engineering technologies.

This book presents state-of-the-art technical contributions based on the most sought-after search algorithms in various domains, including computation intelligence, medicine, e-Commerce, and many others. It consists of a selection of the best contributions from researchers, academicians, and experts on search optimization who present their latest findings and discuss the past, present, and future of this exciting field.

I would like to convey our appreciation to all contributing authors for their excellent chapters. I owe special thanks to Author Service Manager Mr. Josip Knapić and Commissioning Editor Ms. Jelena Germuth at IntechOpen for their kind support and great efforts in bringing the book to fruition. In addition, I also appreciate all those who assisted me throughout the publication process.

Dinesh G. Harkut
Head and Associate Professor,
Department of Computer Science and Engineering,
Prof Ram Meghe College of Engineering and Management,
Amravati, M.S., India

Chapter 1

Introductory Chapter: Search Algorithm - Essence of Optimization

Dinesh G. Harkut

1. Introduction

Every time you hit the search button on Google, the search engine sifts through thousands of searches, if not millions of web pages, to spit out the content we are seeking in a fraction of a second. It is an algorithm – a set of mathematical rules embedded in the software, which makes all this possible. In fact, every time we enlist for a unique identity social security/Aadhaar number, access an automated teller machine, book train or air tickets, or buy merchandise online, we are indirectly expanding the scope and range of algorithms, a mathematical concept whose roots date back to 600 AD with the invention of the decimal system.

Algorithms are nothing but the logically group of instructions aimed at solving a problem or completing a task. Recipes are algorithms like math equations. Computer code is algorithmic. Algorithms are aimed at optimizing everything. Mathematical algorithms include fundamental methods from arithmetic and numerical analysis, which in turn manipulate the data through addition, multiplication of integers, polynomials, and matrices, and may be used for solving a large variety of mathematical problems which arise in many contexts: solution of simultaneous equations, data fitting and integration, and random number generation. Here, the main emphasis is on algorithmic aspects of the methods, rather than the mathematical basis.

Whenever we use a computer, laptop, phone, or a mileage calculator in a car, we are using algorithms, and we may call it programs, or software packages, or apps. They can make things easier, save lives, and surmount disorder. To discuss the effects of technology-enabled assistance in human lives, algorithms are a useful artifact to begin with. Algorithms have penetrated in every aspect of human life and provide a better standard against which to compare human cognition itself. It becomes the new arbiters of human decision-making in almost any area we can imagine like which movie to watch to which house to buy to self-driving cars. Biometrics refer to identifying human being by certain physical features like fingerprints or iris scans. Biometrics-based social security identity card or Aadhaar card, in Indian context which evolved as the India's universal identity card, in turn uses an algorithm to store and retrieve fingerprints and iris scans. Computer scientists have devised algorithms that can analyze a given thumbprint and match it against a database. Because of the overdependence of human beings on computer, indirectly only the algorithms determine whether one gets a job or one get into college or get an apartment; moreover, their work goes largely unnoticed. Algorithms are behind many routines works, but they are still significant decision-making tools in everyone's life.

Deloitte Global predicted that, in time to come, more than 80% of the world's largest software enterprise companies will have cognitive technologies, mediated by algorithms integrated into their products. Algorithms with the perseverance and ubiquity of insects shall automate the processes that are used to require human intervention and rational. These can now achieve basic processes of measuring, monitoring, seeing, or even counting. Our vehicle can guide us where to slow down. Our television sets can now advocate which movies to watch. A grocery can recommend a healthy amalgamation of foods and vegetables for lunch/dinner. Alexa/Siri reminds us important events or anniversaries of dear ones. The overall impact of pervasive algorithms is very hard to calculate because the presence of algorithms in every walk of life, everyday processes, and transactions is now so great and is mostly hidden from public view.

Algorithms are making enormously significant pronouncements in our society in almost every walk of life, ranging from welfare benefits to medicine to transportation to criminal justice and beyond. The ever-increasing assortment and investigation of data and the resulting application of this information can decrease poverty, cure diseases effectively, bring apt resolutions to mankind, places where need is utmost, and dispel epochs of prejudgment, illogical suppositions, vicious practice, and obliviousness of all kinds. Algorithms are now redefining how we think, what we know, and what we think. Algorithms are a black box and are invisible pieces of code that tell a computer how to accomplish a specific task. An algorithm directs the computer what to do in order to produce a certain desired outcome. Every time you do search on internet through any search engine like Google or look at your Facebook feeds or use GPS navigation in your car, you are directly or indirectly interacting with an algorithm. Individuals often demonstrate greater trust on assistance from algorithms compared to non-algorithmic assistance, displaying algorithmic obligation. Counting on algorithms for analytical tasks is typically beneficial. Even simple algorithms, such as weighting all variables equally, can outclass humanoid prediction. Algorithms have begun to intrude on tasks conventionally earmarked for human judgment and are progressively proficient of performing well in innovative and tough tasks. Moreover, at the same time, societal impact, through social media, personal networks, or online assessments and reviews, is one of the most compelling forces affecting individual decision-making.

In short, algorithms are the core entity of the internet, and they manage and run the internet and all online activities like financial transactions, crypto/stock trading, searching, customized browsing, data manipulation, etc. Email knows the destination address and thus knows where to go thanks to the underlying algorithms. Moreover, smartphone mobile apps are nothing but algorithms. Computer and video games are algorithmic storytelling. Book or movie recommendation, online dating, leisure\travel web portals, and so on would not function properly and efficiently without algorithms. Artificial intelligence (AI) is nothing but algorithms. GPS mapping systems make use of algorithms to get entity from location X to location Y. Every single piece of the object people see on social media is brought to them by means of algorithms. Moreover, everything people do in everyday life and see on the web is an outcome of some algorithm or other. Every time we sort or arrange data in a worksheet, algorithms plays vital role, and almost every financial transaction is accomplished today by algorithms. Algorithms make every electronic gadget to respond to voice commands, organize and sort photos, recognize and identify faces, and build and drive automobile cars. Hacking, cyberattacks, and cryptographic code-breaking exploit algorithms to the next level. Algorithms are often sophisticated, elegant, and amazingly useful tools used to accomplish various categories of tasks. They are mostly hidden and invisible aids, enhancing and augmenting human lives in increasingly efficient ways. Algorithms will continue to face the ever-increasing impact over the next few decades, influencing people's work and personal lives and the ways they interact with information, organizations like health care service providers, not-for-profit/government institutions, banking/financial sector, retailers/traders, education, media corporates houses and entertainment industry, and each other. The hope is that algorithms will help people swiftly and impartially perform the tasks and get the desired products, information, and services. The major apprehension is that algorithms can deliberately or unconsciously create discrimination and thus enable social engineering to create biased narrative and have other harmful societal impacts.

The term algorithm which finds application in computer science is universally used to describe problem-solving methods that help for the implementation of computer programs. Mostly, algorithms involve complicated methods of manipulating and organizing the data involved in the computation. Basically, they involved the manipulation of data based on some mathematical model and which typically find applications of: searching, sorting, string processing, geometric algorithms, graph algorithms, genetic algorithms (GAs), neural network, etc.

2. Searching

Searching is a method for finding certain things in given data/files which are of vital importance. There are different categories of search methods: basic and advanced, like one using trees and digital key transformations, including balanced trees, hashing binary search trees, and digital search trees, and trying different methods appropriate for different types of files. These methods in turn are related to each other and possess much resemblance with sorting methods.

3. Sorting

Sorting is a method of rearranging files in the order that are covered in some depth due to their fundamental importance. A large variety of methods were developed, described, and compared. Algorithms including priority queues, selection, merging, and several related problems are created. Some of these algorithms are used as the basis for other more complex algorithms.

4. String processing

String processing algorithms include a range of methods for dealing with (long) sequences of characters. String searching basically leads to pattern matching which in turn leads to parsing. File compression techniques and cryptology are also part of advanced string processing applications.

5. Geometric algorithms

Geometric algorithms encompass an assortment of procedures for resolving problems by involving points and lines along with other simple geometric objects. Moreover, it also finds application of ideas from several mathematical disciplines like algebra, combinatorics, topology, and differential geometry. During the last two decades, most of the geometric applications like CAD/CAM, computer graphics, VLSI design, molecular biology, robotics, GIS, spatial databases, sensor networks, machine learning, and scientific computing become the motivation for computational geometry to evolve as a full-fledged discipline of theoretical computer science. Some typical high-end applications of geometric algorithms includes: optimal airspace design, air traffic controller's traffic balancing and automatic sectorization of airspace, wireless sensor networks design, analysis of 2D electrophoresis gels, prediction of resilience, and recovery of damage in neural networks.

6. Graph algorithms

Graphs algorithms are the natural way to understand connection between the linked data and thus reveal the relationships in data. Exploring and tracking these interlinking connections divulge new insights and influence and facilitate to analyze each data point as part of a bigger picture. Graph algorithms are useful in a variety of complex difficult structures and reveal difficult-to-find patterns ranging from finding bottlenecks, susceptibilities to detect communities, fraud rings, improving machine learning predictions to predict the spread of disease or ideas, and thus enable us to leverage relationship within data to devise intelligent solution to enhance the effectiveness of machine learning models. Graph algorithms types such as exact or approximated, static or dynamic, distributed or centralized, deterministic or randomized, and matching and network flow, minimal spanning tree, and shortest path are some of the general approaches developed for searching in the graph.

7. Genetic algorithms

A genetic algorithm (GA) is a search-based heuristic algorithm used for solving optimization problems in machine learning that is based on genetics and natural selection. GA is a subset of evolutionary algorithms which is based on behavior of chromosomes and their genetic structures which uses evolutionary generational cycle starting from initialization, fitness assignment, selection, reproduction, replacement, and termination, to yield high-quality solutions. GA makes use of various operations which enhance or replace the population to deliver a better-quality suitable solution and get inspiration from evolution and natural selection. Through the process of natural selection, organisms regulate to augment their likelihoods for endurance in a given situation. Eventually, incompetent elements perish from the population, to be replaced by successful-solution descendants. Apart from the applications in multimodal optimization, DNA analysis, design of aircraft, and genetic algorithms are found to be efficient and cost-effective plan for tricky traveling salesman problem. It does not need derivative information and possess excellent parallel capabilities which refine and optimize the solution to the multiobjective problems, and discrete and continuous functions. The idea behind genetic algorithms is extremely alluring.

8. Neural networks

Neural networks are basically inspired by the biological nervous system or neural networks in the brain and basically parallel computing devices. It is one of the most emerging areas in data science which revolutionizes and eventually leads to tremendous growth of artificial intelligence, machine learning, and deep learning and basically designed to recognize patterns and extracting features that can be fed to other relevant algorithms for further classifications and clustering. The elementary computational unit of a neural network is a neuron also called as node, and each node or neuron is linked with certain weight. This weight is assigned as per the relative importance of that particular neuron or node. Neural networks are simply weighted digraphs with neurons acting as vertices, and weights on edges denote the connection strength of the pair. Each node or neuron collects values or weights from other neurons and accordingly computes the output.

Gradient Decent, Conjugate Gradient, Newton's Method, and Quasi-Newton Method are some of the popular optimization algorithms which find applications in context with neural network. Speed and memory footprint of all these algorithms may vary, but an ultimate objective is to accomplish various intricate computational task faster than the traditional systems.

In neural network, training of the neurons is accomplished by appropriately modifying connection strength in response to training data. Most apt application of neural networks is forecasting and classification of applications, such as gene predication, optical character recognition, and stock market time series prediction.

Having discussed about all these different types of algorithms, one thing which is most apparent that, irrespective of types of algorithms, searching lies at the heart of all. Rather, it is the intrinsic part of all algorithms as it finds application one way or other, and it is one of the first things any algorithm designer should try in the quest for efficiency. Basically, sorting is directly based on searching, wherein we use to compare, i.e., search for specific pattern of string for comparison, and then rearrange those strings based on desired sorting pattern of ascending or descending. Searching can be used to illustrate most algorithm design paradigms. Data structure techniques, divide-and-conquer, randomization, and incremental construction all lead to popular sorting algorithms. Binary search and its variants are the essential divide-and-conquer algorithms. Depth-first and breadth-first search provide mechanisms to visit each edge and vertex of the given graph. Strategy represents the pursuit for the big picture, the framework around which we construct our path to the goal. Strategies are used to win the minor skirmishes we must fight along the way. They prove the basis of most simple and efficient graph algorithms. Simulated annealing is a simple and effective technique to efficiently obtain good but not optimal solutions to combinatorial search problems. Combinatorial search, improved with tree pruning techniques, can then be used to find the more optimal solution of small optimization problems. Ingenious pruning procedures can speed up combinatorial search to a remarkable extent. Proper trimming will have a greater impact on search time than any other factor. Historically, computers have spent more time searching and sorting than doing anything else all put together. A quarter of all mainframe cycles are spent in searching and sorting data. Although it is unclear whether this remains true on smaller computers, but still searching and sorting remain the most ubiquitous combinatorial algorithm problem in practice.

An important key to algorithm design is to use searching/sorting as a basic building block, because once a set of items is sorted, many other problems become easy. Consider the few of the applications like:

- **Closest pair** – In a given a set of *n* different numbers, suppose we are interested in finding the pairs of numbers having minimum or no deviation or minimal difference between them. One way to find the desired pairs in quickest possible way is to arrange these numbers in sorted order either in ascending or descending order. Once sorted, the closet pair of numbers shall spatially reside next to each other. Thus, merely a linear scan through the sorted list will complete the job.

- **Element uniqueness** – In a given set of n items or list of numbers, suppose we are interested in finding the occurrence of any duplicates, the most efficient and appropriate algorithm is to sort these item lists followed by simple linear scan. By verifying and checking, indirectly searching all adjacent pairs, duplicate item can be pointed out. This can be treated as one of the special cases of the closest-pair problem, wherein instead of finding minimal deviation, we are typically interested in finding the elements in the list with deviation or gap of difference of zero.

- **Frequency distribution** – In a given a set of n items, suppose we want to find the major number of times or count of appearance/occurrence of particular element, i.e., frequency of occurrence, arranged the given list in sorted order either in ascending or descending, scan the list from left to right and go on counting them, as all matching elements will be lumped together during sorting. To find out how frequently an arbitrary selected element k occurs, one can start by looking up k using binary search in a sorted array of keys. By scanning the list from left of this point until the element is not k and then moving to the right, we can find this count c in linear time, where c is the number of occurrences of k. The number of instances of k can be found in time by using binary search to look for the positions of both and where it is arbitrarily small, and then by taking the difference of these positions.

- **Selection** – In a given list of numbers in array, suppose one is interested in finding the kth biggest or largest item, the desired k^{th} largest can be found in fix constant time. If the numbers in the arrays are arranged and placed in proper sorted order, we simply need to look for the kth position in the array, as the median element appears in the $(n/2)^{nd}$ position in sorted order list.

- **Convex hulls** – In a given set of n points in two dimensions, suppose we want to find the polygon having smallest area such that it encompasses all the points. Here, the convex hull is which is just like a rubber band stretched over the points in the given plane and then released gives a nice representation of the shape of the points and is the most important building block for more sophisticated geometric algorithms. Moreover, if the given points are arranged in sorted order in either of the coordinate, the points can be inserted either from left to right or bottom to top, respectively, into the hull. Consider the case that points are sorted on x coordinate, as the rightmost point is always lies on the edge, and it will be inserted into the hull. However, adding new rightmost point may cause others point to be deleted, and we can swiftly identify such points as they fall inside the polygon formed by adding these new points. Thus, points to be deleted are the neighbors of the preceding point we injected, so they will be easy to identy. The total time is linear after the sorting has been done. Although some of these issues can be solved in linear time using more sophisticated algorithms, sorting which is incidentally based on searching at core provides quick and easy solutions to all of these problems. It is a rare application whose time complexity is such that sorting proves to be the bottleneck, especially a bottleneck that could have otherwise been removed using clever algorithmics.

Thus, searching/sorting is the most thoroughly studied problem in computer science. Literally, loads of diverse algorithms are known, most of which have some advantages over all other algorithms in certain circumstances. Most of the stimulating concepts used in the design of algorithms appear in the context of searching/sorting such as data structures, divide-and-conquer, and randomized algorithms.

Searching/sorting is a natural laboratory for studying basic algorithm design paradigms, since many useful techniques lead to interesting searching/sorting algorithms. Searching/sorting thus becomes the basic building block around which many other algorithms are built. By understanding searching thoroughly and devising effective searching techniques, we can obtain an amazing amount of power to solve other problems and thus will become the essence for optimizations. ·

Author details

Dinesh G. Harkut
Prof. Ram Meghe College of Engineering and Management Badnera, India

*Address all correspondence to: dg.harkut@gmail.com

IntechOpen

Chapter 2

Search Algorithms on Logistic and Manufacturing Problems

Gladys Bonilla-Enriquez
and Santiago-Omar Caballero-Morales

Abstract

The supply chain comprehensively considers problems with different levels of complexity. Nowadays, design of distribution networks and production scheduling are some of the most complex problems in logistics. It is widely known that large problems cannot be solved through exact methods. Also, specific optimization software is frequently needed. To overcome this situation, the development and application of search algorithms have been proposed to obtain approximate solutions to large problems within reasonable time. In this context, the present chapter describes the development of Genetic Algorithms (an evolutionary search algorithm) for vehicle routing, product selection, and production scheduling problems within the supply chain. These algorithms were evaluated by using well-known test instances. The advances of this work provide the general discussions associated to designing these search algorithms for logistics problems.

Keywords: vehicle routing problem, knapsack problem, flow-shop Scheduling, local-search Algorithms, genetic algorithms

1. Introduction

According to the Council of Supply Chain Management Professionals (CSCMP), logistics is defined as the process of planning, implementing and controlling all operations and information flow for the efficient and effective transportation and storage of goods or services from a point of origin to a point of consumption. As presented in **Figure 1**, many operations are involved in a logistics network, and manufacturing is a crucial operation to transform inbound goods (e.g., raw materials) into outbound goods (e.g., end products, sub-assemblies, work-in-process, etc.) throughout this network.

Due to the complexity of these operations, where many of them involve problems of NP-hard computational complexity, research and improvement efforts require the use of advanced of quantitative and qualitative strategies and tools. Among these, the use of Search Algorithms such as meta-heuristics has been proposed to solve to near-optimality large NP-hard problems within reasonable time [1].

As presented in **Figure 1**, transportation is needed for the efficient flow of goods throughout the supply chain (SC). Thus, the analysis and solution of routing problems are the first set of problems to be addressed in this chapter.

Then, manufacturing planning is needed to achieve the required quantities of sub-assemblies and end-products to supply the customers (or even other suppliers)

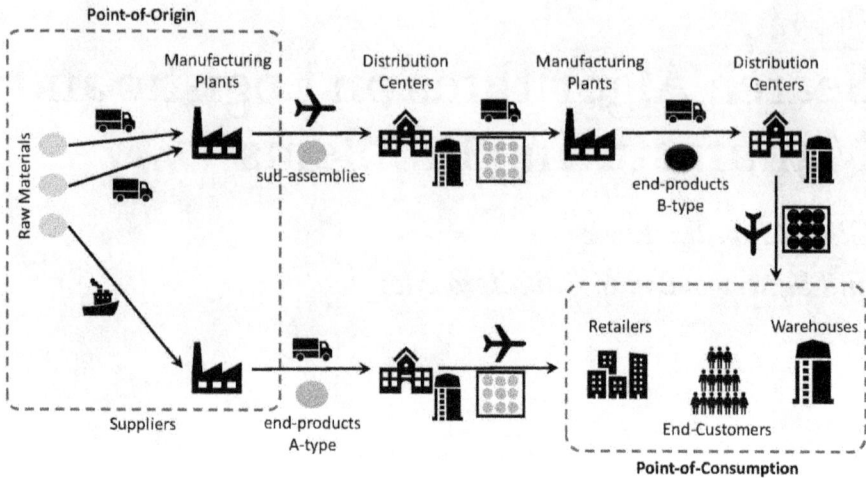

Figure 1.
General example of a logistics network.

in time through the SC. Thus, production planning problems are the second set of problems to be addressed in this chapter. Note that both sets are mutually important and dependent for the appropriate performance of the SC.

While there are many search algorithms or meta-heuristic approaches to solve these problems, this chapter addresses the specific configuration settings to apply Genetic Algorithms (GA) to solve both sets of problems. As the solutions have different representations (i.e., permutations, binary chains, real numbers), having a common algorithmic base can lead to a better understanding for successful implementation for other problems and contexts.

GA are based on the principle of natural selection of "survival of the fittest" where individuals within a population compete between each other for vital resources (i.e., food, shelter, etc.) and/or to attract mates for reproduction. Due to this selection mechanism, it is expected that poorly performing individuals have less chance to survive in contrast to the most adapted or "fit" individuals which are more likely to reproduce, inheriting their good characteristics to their offspring to make them better and more adapted to their environment [2].

Figure 2 presents the general structure and main elements of a GA. This meta-heuristic is population-based. Thus, it works by continuously improving on a set of solutions by using reproduction operators which facilitate the search mechanisms for the solution space of the problem. This set, known as the population, consists of N feasible solutions which are evaluated through a fitness function (i.e., the total distance equation, or objective function, to determine the total cost associated to each solution). Then, the solutions with the best fitness values become candidates for reproduction to (hopefully) inherit their best features to new solutions and improve the overall population in the next generation (iteration). It is expected that after X generations the mean fitness of the population converges to a local optimum.

Within this context, the present chapter addresses the different representations of candidate solutions, fitness functions, and reproduction operators, for the application of GA to solve the following sets of problems:

- Routing Planning (Section X.2): Traveling Salesman Problem (TSP) and Capacitated Vehicle Routing Problem (CVRP).

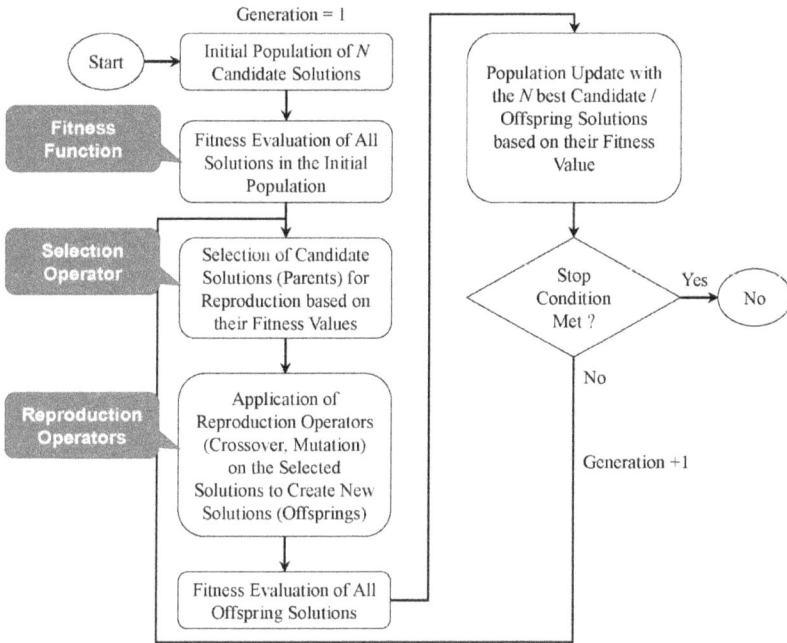

Figure 2.
General structure and main elements of a GA.

- Production and Selection of the Most Profitable Goods (Section X.3): Economic Lot Problem with Multiple Items and Knapsack Problem.

- Production Scheduling (Section X.4): Permutation Flow-Shop Scheduling Problem.

This chapter ends with a discussion of the results and the practical implications of the future work (Section X.5).

2. Genetic algorithm for route planning problems

2.1 Traveling salesman problem

The Traveling Salesman Problem (TSP) represents the scenario of a salesperson who must visit each place within a set of cities or towns. This must be performed with the following considerations: the salesperson starts and ends the whole journey at a single location (i.e., the main office) and must visit each place only once [3]. Although this is the basic understanding of the TSP, the main feature of finding a single route, or sequence of minimum distance or cost, is shared by other real-world applications such as vehicle routing [4], production planning [5], service time [6], and design of computer networks [7]. **Figure 3** presents an overview of the TSP model with $n = 12$ cities.

Note that each single route that complies with the previous restrictions represents a candidate solution, and there are as much as $n!$ candidate solutions if brute search were to be considered as solving method to find the optimal or best solution. Just for the example presented in **Figure 3**, there are up to $12! = 479'001,600$ or

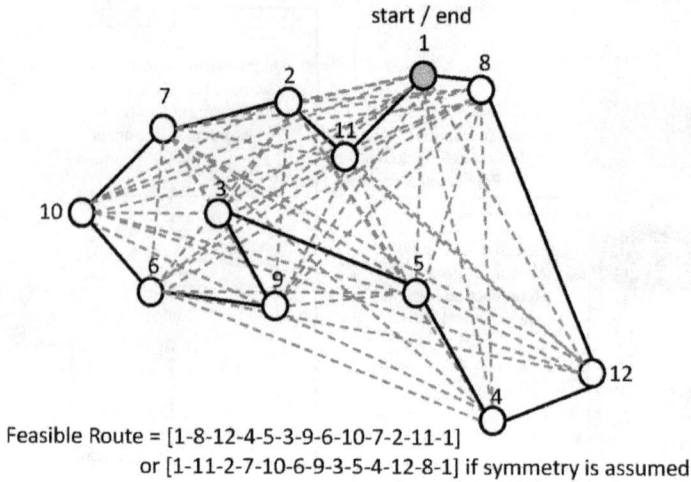

Figure 3.
Example of a feasible TSP solution with n = 12 cities.

479.00e+006 feasible solutions to visit all 12 cities. This number increases exponentially as n increases linearly. Thus, if just a single city is added to the TSP problem, the number of feasible solutions can increase to 13! = 6.23e+009.

This leads to a problem with an infinite solution space if large sets of cities are considered. This classifies the TSP as an NP-hard problem, which is very difficult to solve within reasonable time, even with the most advanced computational systems. Thus, different meta-heuristics have been developed to provide fast near-to-optimal solutions. Among these meta-heuristics the following can be mentioned [8]: Nearest Neighbor (NN), Simulated Annealing (SA), Tabu Search (TS), Genetic Algorithm (GA), Ant Colony Optimization (ACO), Particle Swarm Optimization (PSO) and Tree Physiology Optimization (TPO).

As presented in [8] GA and SA are among the most suitable heuristics, achieving error gaps from best known solutions within the 10% mark for small ($n < 100$), moderate ($100 < n < 150$) and large ($150 < n < 450$) TSP instances. However, within the context of TSP solutions, it is always recommended to test the solving methods with very large instances (i.e., n > 500) to corroborate their performance.

Thus, the developed GA considers TSP instances with $n \approx 1000$. For this purpose, the GA considers the structure presented in **Figure 2** with the settings and reproduction operators presented in **Table 1** and described in **Figure 4**.

Parameter	Setting
Generations (Iterations)	1000
Fitness Function	Symmetric Euclidean Distance of TSP Route
Population Size	According to the size of the TSP
Selection	Tournament
Reproduction Operators:	
Crossover	Partially-Matched Crossover (PMX)
Mutation	Swap, Inversion

Table 1.
GA settings for the TSP.

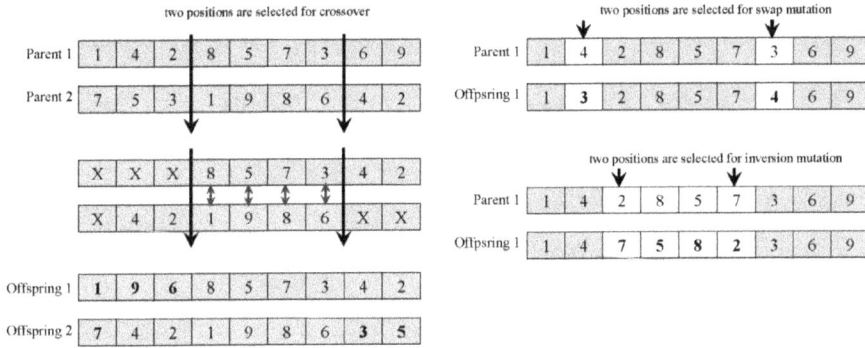

Figure 4.
Partially-matched crossover, and swap/inversion mutation operators for the TSP.

Implementation of the GA was performed in MATLAB with an Intel Core
i7–5500 CPU at 2.40 GHz and 8GB RAM. Testing was performed with a set of TSP
instances from the TSPLIB95 database [9]. The details of these instances, including
the GA's population size N used for each case, are presented in **Table 2**. The results
of the tests can be observed in **Figure 5** and **Figure 6**.

As presented in **Figure 5**, the mean error gap through all instances begins to
decrease as the selection and reproduction mechanisms of the GA start to operate
on the initial and updated populations. By the 300th generation the mean error gap
decreases under the 10% mark to finally reach an approximate of 7% by the 1000th
generation. This corroborates the performance reported in [8].

Finally, **Figure 6** presents the performance of the GA based on the size of the
test instances (n). With the settings reported in **Table 1**, as n increases, the GA takes
more time to converge to a local optimum which, in some cases, it is slightly over the
10% mark. Also, the size of the population (N) must be increased to improve the
search performance.

Based on these findings, particularly for the TSP with $n \approx 1000$, the following
recommendations can be made:

- Diversification of solutions depends of the size of the population (N) and the
TSP (n). Because N is the only controllable parameter, it is important to find
an appropriate balance between it and n because a large N can increase the
computational memory load of the algorithm which is already affected by n.

- A larger number of generations should be considered for large TSP problems.
This because convergence may get slower due to n, independently of the
reproduction or selection operators, or the size of the population.

- Integration with other heuristics or meta-heuristics can improve on the initial
population or some of the search operators, and thus, on the convergence of
the GA through all generations. This process, called hybridization, has led to
obtain very suitable results for large TSP instances [10].

As an example of hybridization, **Figures 5** and **6** present the performance of
the revised GA (hybrid-GA) with a much smaller N (= 50 for all instances) and a
Greedy algorithm to improve four offspring (two by crossover, one by flip mutation, one by swap mutation) which are included within the updated population.
This increases the speed of the GA, reaching the 10% by the 100th generation, with

a final mean error gap of 5% by the 1000th generation. Also, improvement of the large instances ($n > 500$) is observed, achieving error gaps under the 10% mark.

N	Size of the TSP (n)	Name of the Instance	N	Size of the TSP (n)	Name of the Instance
200	51	eil51.tsp	200	198	d198.tsp
200	52	berlin52.tsp	300	200	kroA200.tsp
200	70	st70.tsp	300	200	kroB200.tsp
200	76	eil76.tsp	300	225	ts225.tsp
200	76	pr76.tsp	300	225	tsp225.tsp
200	99	rat99.tsp	300	226	pr226.tsp
200	100	kroA100.tsp	300	262	gil262.tsp
200	100	kroB100.tsp	300	264	pr264.tsp
200	100	kroC100.tsp	300	280	a280.tsp
200	100	kroD100.tsp	300	299	pr299.tsp
200	100	kroE100.tsp	300	318	lin318.tsp
200	100	rd100.tsp	500	400	rd400.tsp
200	105	lin105.tsp	500	417	fl417.tsp
200	107	pr107.tsp	500	439	pr439.tsp
200	124	pr124.tsp	500	442	pcb442.tsp
200	127	bier127.tsp	500	493	d493.tsp
200	130	ch130.tsp	500	574	u574.tsp
200	136	pr136.tsp	500	575	rat575.tsp
200	144	pr144.tsp	800	654	p654.tsp
200	150	ch150.tsp	800	657	d657.tsp
200	150	kroA150.tsp	800	724	u724.tsp
200	150	kroB150.tsp	1500	783	rat783.tsp
200	152	pr152.tsp	1500	1002	pr1002.tsp
200	159	u159.tsp	1500	1060	u1060.tsp
200	195	rat195.tsp	1500	1084	vm1084.tsp

Table 2.
TSPLIB instances for GA testing.

2.2 Capacitated vehicle routing problem

The Capacitated Vehicle Routing Problem (CVRP) represents an extension on the TSP. As shown in **Figure 7**, the CVRP determines a set of routes that start and end at a specific place or location (e.g., a distribution center). These routes must visit or serve a finite number of locations and meet their demand requirements. Each route must be served by a single vehicle (e.g., a salesperson) with finite capacity, and only one vehicle can serve a location. Thus, the CVRP can be understood as a variant of the multiple-TSP with capacity restrictions [11].

As in the case of the TSP, the CVRP is a combinatorial problem of NP-hard complexity which cannot be solved within a reasonable polynomial time [12]. Due to this, the CVRP has been addressed by different meta-heuristics such as Tabu -Search (TS) [13, 14], GA [15], SA [16, 17], and Particle Swarm Optimization (PSO) [18].

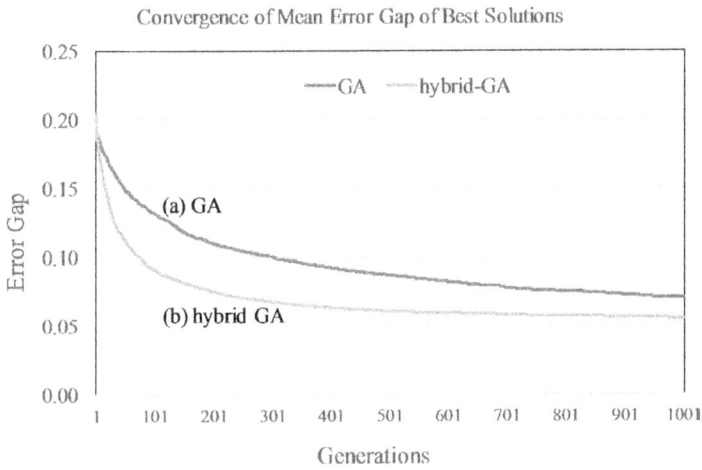

Figure 5.
Mean error gap across all TSP test instances with (a) the GA, and (b) the revised hybrid-GA.

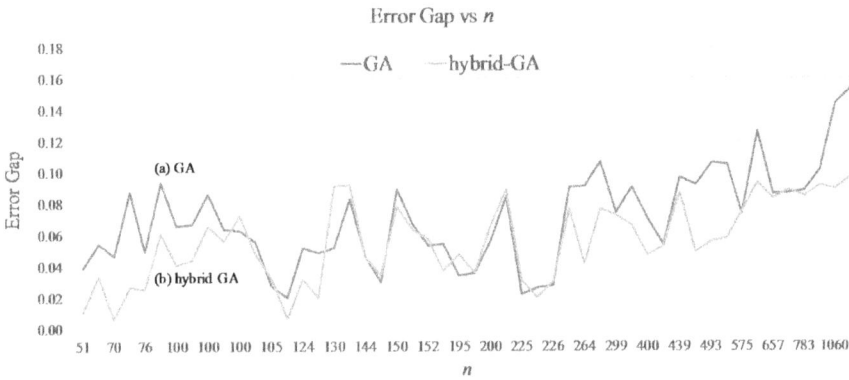

Figure 6.
Error gap across all TSP test instances with (a) the GA, and (b) the revised hybrid-GA.

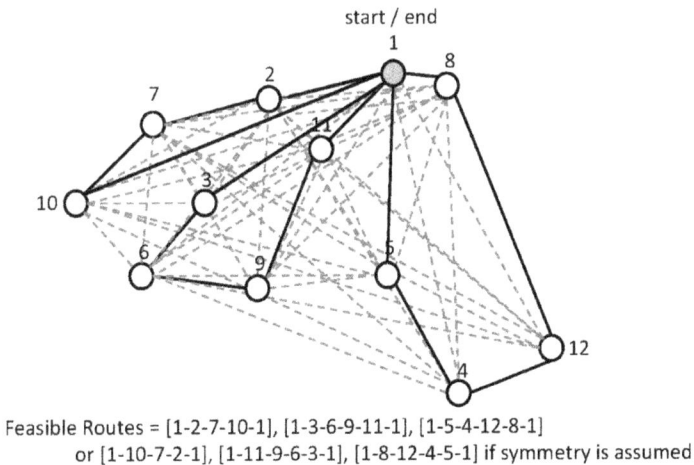

Feasible Routes = [1-2-7-10-1], [1-3-6-9-11-1], [1-5-4-12-8-1]
or [1-10-7-2-1], [1-11-9-6-3-1], [1-8-12-4-5-1] if symmetry is assumed

Figure 7.
Example of a feasible CVRP solution with n = 12 cities and 3 routes.

For this case, the GA presented in **Figure 2** was modified to solve the CVRP. The GA and its configuration settings are presented in **Figure 8** and **Table 3** respectively. Note that the reproduction operators remain the same as considered for the TSP. Testing was performed with a set of instances from the CVRPLIB database [19, 20]. **Table 4** presents the details of the selected instances.

As presented in **Figure 9**, the mean error gap reaches the 10% mark by the 200th generation, with an approximate of 8.5% by the 1000th generation. In contrast to the patterns observed in **Figure 6**, in **Figure 10** there is not a clear relationship between the size of the instance (n) and the error gap. Thus, there are large instances with very small error gaps (approximately 6%) and medium instances with large error gaps (over 10%). This however is expected because there are more tasks to be performed on the CVRP such as route segmenting and capacity restriction compliance. This leads to frequently consider GAs for small CVRP instances ($n < 200$) [15, 21].

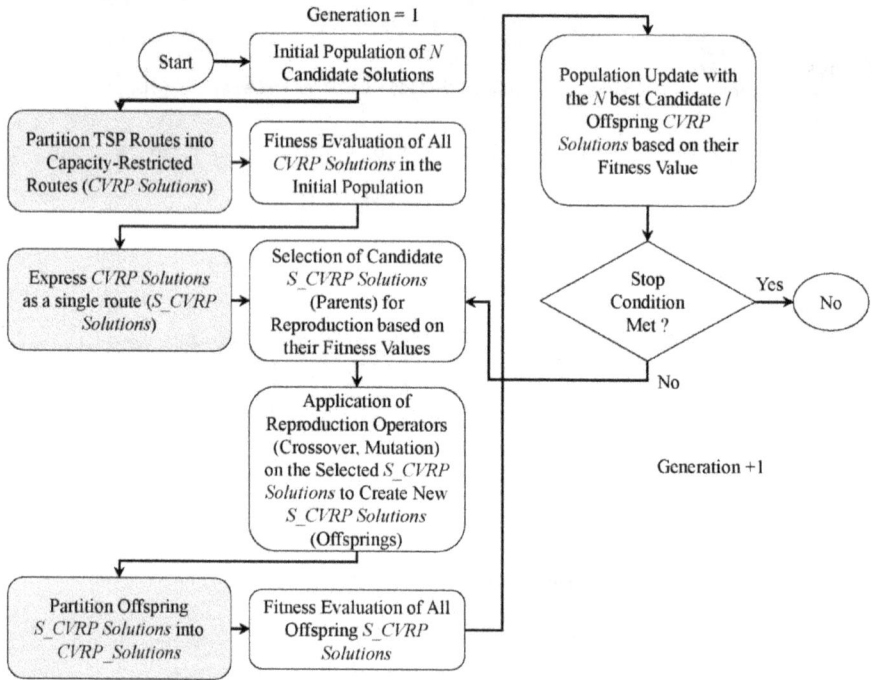

Figure 8.
Modified structure of the GA for the CVRP.

Parameter	Setting
Generations (Iterations)	1000
Fitness Function	Symmetric Euclidean Distance of CVRP Routes
Population Size	$N = 100$
Selection	Tournament
Reproduction Operators:	
Crossover	Partially-Matched Crossover (PMX)
Mutation	Swap, Inversion

Table 3.
GA settings for the CVRP.

Size of the CVRP (n)	Number of CVRP Routes	Name of the Instance	Size of the CVRP (n)	Number of CVRP Routes	Name of the Instance
100	25	X-n101-k25	335	84	X-n336-k84
105	14	X-n106-k14	343	43	X-n344-k43
109	13	X-n110-k13	350	40	X-n351-k40
114	10	X-n115-k10	358	29	X-n359-k29
119	6	X-n120-k6	366	17	X-n367-k17
124	30	X-n125-k30	375	94	X-n376-k94
128	18	X-n129-k18	383	52	X-n384-k52
133	13	X-n134-k13	392	38	X-n393-k38
138	10	X-n139-k10	400	29	X-n401-k29
142	7	X-n143-k7	410	19	X-n411-k19
147	46	X-n148-k46	419	130	X-n420-k130
152	22	X-n153-k22	428	61	X-n429-k61
156	13	X-n157-k13	438	37	X-n439-k37
161	11	X-n162-k11	448	29	X-n449-k29
166	10	X-n167-k10	458	26	X-n459-k26
171	51	X-n172-k51	468	138	X-n469-k138
175	26	X-n176-k26	479	70	X-n480-k70
180	23	X-n181-k23	490	59	X-n491-k59
185	15	X-n186-k15	501	39	X-n502-k39
189	8	X-n190-k8	512	21	X-n513-k21
194	51	X-n195-k51	523	137	X-n524-k153
199	36	X-n200-k36	535	96	X-n536-k96
203	19	X-n204-k19	547	50	X-n548-k50
208	16	X-n209-k16	560	42	X-n561-k42
213	11	X-n214-k11	572	30	X-n573-k30
218	73	X-n219-k73	585	159	X-n586-k159
222	34	X-n223-k34	598	92	X-n599-k92
227	23	X-n228-k23	612	62	X-n613-k62
232	16	X-n233-k16	626	43	X-n627-k43
236	14	X-n237-k14	640	35	X-n641-k35
241	48	X-n242-k48	654	131	X-n655-k131
246	47	X-n247-k50	669	126	X-n670-k130
250	28	X-n251-k28	684	75	X-n685-k75
255	16	X-n256-k16	700	44	X-n701-k44
260	13	X-n261-k13	715	35	X-n716-k35
265	58	X-n266-k58	732	159	X-n733-k159
269	35	X-n270-k35	748	98	X-n749-k98
274	28	X-n275-k28	765	71	X-n766-k71
279	17	X-n280-k17	782	48	X-n783-k48
283	15	X-n284-k15	800	40	X-n801-k40
288	60	X-n289-k60	818	171	X-n819-k171
293	50	X-n294-k50	836	142	X-n837-k142
297	31	X-n298-k31	855	95	X-n856-k95

Size of the CVRP (n)	Number of CVRP Routes	Name of the Instance	Size of the CVRP (n)	Number of CVRP Routes	Name of the Instance
302	21	X-n303-k21	875	59	X-n876-k59
307	13	X-n308-k13	894	37	X-n895-k37
312	71	X-n313-k71	915	207	X-n916-k207
316	53	X-n317-k53	935	151	X-n936-k151
321	28	X-n322-k28	956	87	X-n957-k87
326	20	X-n327-k20	978	58	X-n979-k58
330	15	X-n331-k15	1000	43	X-n1001-k43

Table 4.
CVRPLIB instances for GA testing.

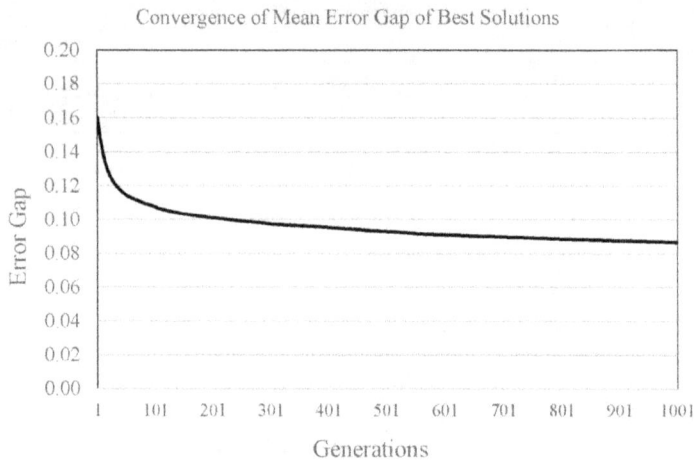

Figure 9.
Mean error gap across all CVRP test instances with the GA.

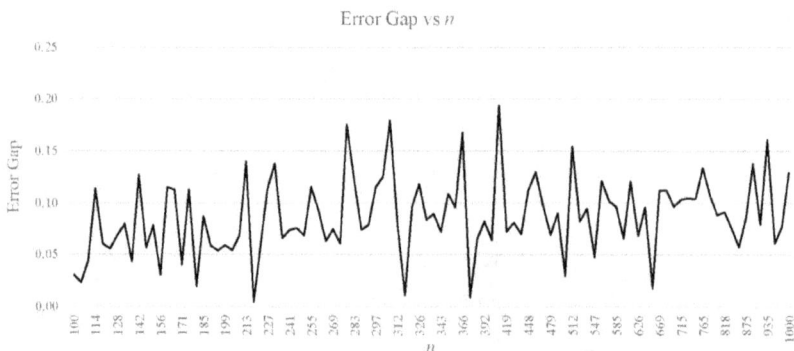

Figure 10.
Error gap across all CVRP test instances with the GA.

Based on these findings, particularly for the CVRP with $n \approx 1000$, the following recommendations can be made:

- Due to the size of the population and the additional tasks, faster processes are needed for diversification of solutions. In example, Tabu Search (TS) uses small sets of candidate solutions (neighbors) through the consideration of movements (or moves). Also, convergence to a local optimum can be minimized by forbidding certain moves (e.g., make them *tabu*) which would make the algorithm to revisit a region within the solution space. This is an advantage when compared to GA, which requires full-candidate solution populations, and avoidance of previously obtained solutions may require additional tasks.

- Hybridization can improve the convergence and overall search performance of near-optimal solutions. In example, implementing a *tabu* mechanism on the population can reduce the rate of previously visited solutions (same solutions) and even dynamically reduce the size of the population.

- Initial convergence of the GA, and overall initial performance, may benefit from an initial population with very suitable solutions. However, this may restrict the diversification of solutions through later generations.

3. Genetic algorithm for production and selection of goods

3.1 Economic lot quantity with multiple items

In manufacturing, an important aspect is the supply of resources such as raw materials, sub-assemblies, end/final products, etc. The availability of these resources must comply with time and cost restrictions.

Within this aspect, the Economic Lot Quantity (EOQ) models are aimed to estimate the lot size Q which minimizes operational costs associated to inventory management. In general, Q minimizes the following cost function:

$$T = \left(\frac{D}{Q}\right)C_o + \left(\frac{Q}{2}\right)C_h. \tag{1}$$

Where C_o is the ordering cost per lot, C_h is the holding cost per unit of product, and D is the cumulative demand through a planning horizon [22]. As presented in **Figure 11**, Q can also be understood as the lot size that equals the total order cost with the total holding cost through a planning horizon (and this leads to minimize T):

$$\left(\frac{D}{Q}\right)C_o = \left(\frac{Q}{2}\right)C_h. \tag{2}$$

Note that Eq. (2) leads to define:

$$Q = \sqrt{\frac{2DC_o}{C_h}}. \tag{3}$$

Which computes the optimal value for Q. Now, if N items with independent orders are considered, then:

$$T = \sum_{i=1}^{N}\left(\left(\frac{D_i}{Q_i}\right)C_{oi} + \left(\frac{Q_i}{2}\right)C_{hi}\right). \tag{4}$$

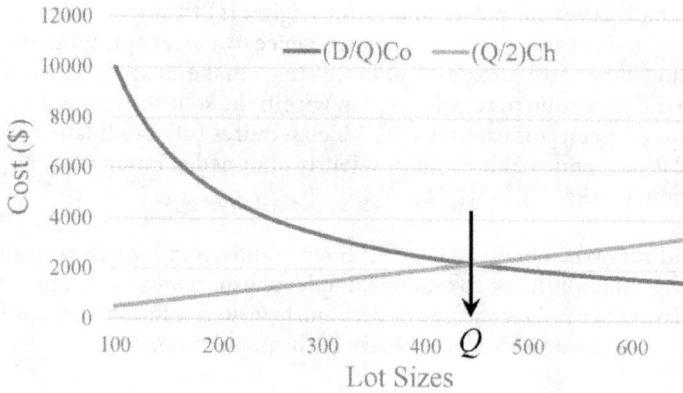

Figure 11.
Inventory management costs associated to the EOQ model.

Under the assumption of independence, Q_i can be optimally computed by using Eq. (3) for each item [22]. Thus, for the present case, the GA is only developed to verify its efficiency to solve the EOQ to optimality with a large N.

The GA follows the standard structure presented in **Figure 2**. As the solution consists of a set of Q_i values, the restrictions associated to permutations (such as in the case of TSP/CVRP) are not present. Thus, a simpler crossover operator can be used.

Figure 12 presents an overview of the linear crossover operator used for the GA. On the other hand, **Table 5** presents the configuration settings of the GA.

The average results for different randomly generated sets of N products are presented in **Figure 13**. As this is a simpler problem than both, the TSP and the CVRP, optimality can be reached within 100–200 generations. Note that it is always recommended to select an exact method if it is available and results can be obtained within very reasonable time.

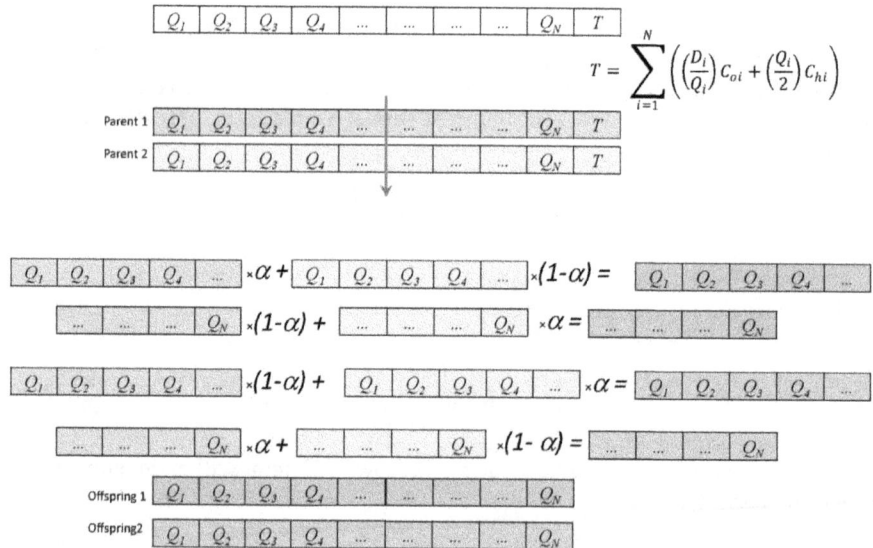

$$T = \sum_{i=1}^{N} \left(\left(\frac{D_i}{Q_i}\right) C_{oi} + \left(\frac{Q_i}{2}\right) C_{hi} \right)$$

Figure 12.
Linear crossover operator for the multiple-item EOQ ($\alpha = 0.5$).

Parameter	Setting
Generations (Iterations)	2000
Fitness Function	Total Inventory Management Cost (T)
Population Size	$N = 1000$
Selection	Tournament
Reproduction Operators:	
Crossover	Linear Crossover
Mutation	Swap, Inversion

Table 5.
GA settings for the multiple-item EOQ.

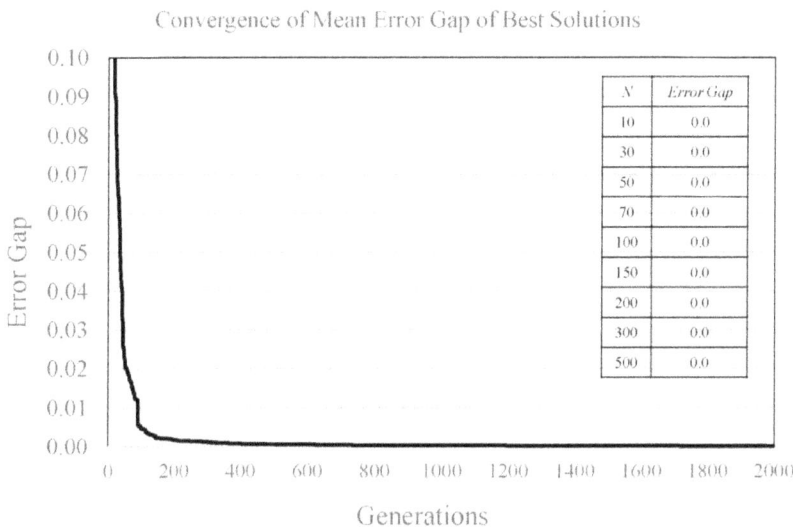

Convergence of Mean Error Gap of Best Solutions

N	Error Gap
10	0.0
30	0.0
50	0.0
70	0.0
100	0.0
150	0.0
200	0.0
300	0.0
500	0.0

Figure 13.
Mean error gap across all multiple-item EOQ with the GA.

3.2 Knapsack problem

The Backpack or Knapsack Problem (KP) is a binary multicriteria problem of NP-hard computational complexity and it is frequently considered as a strategy to select items to maximize profits without affecting capacity restrictions [23, 24].

The KP can be mathematically formulated as a vector of binary variables x_j where $x_j = 1$ if the item j is selected, and $x_j = 0$ otherwise. Then, if p_j is a measure of importance (in this case, profit) for an item j, w_j represents the size of said item, and cv is the size of the backpack, the problem refers to the selection of the quantity of all elements whose binary vectors x_j satisfy the following restrictions [24]:

$$\sum_{j=1}^{n} w_j x_j \leq cv \tag{5}$$

$$x_j \in \{0,1\}, j = 1,\ldots,n \tag{6}$$

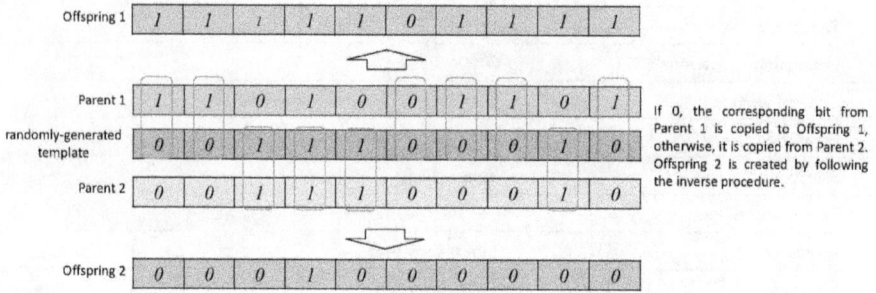

Figure 14.
Uniform crossover operator for the KP.

Parameter	Setting
Generations (Iterations)	100
Fitness Function	Total Profit of Selected Goods (T)
Population Size	$N = 1000$
Selection	Tournament
Reproduction Operators:	
Crossover	Uniform Crossover
Mutation	Swap, Inversion

Table 6.
GA settings for the KP.

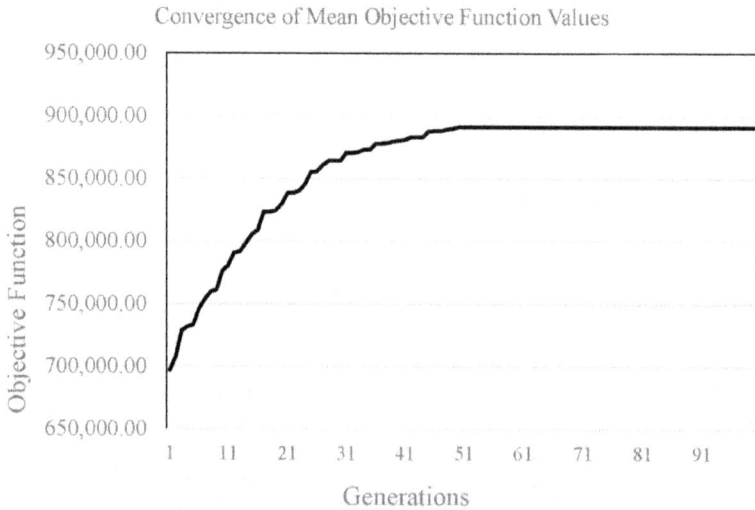

Figure 15.
Mean objective function values across all KP instances with the GA.

These must contribute to maximize the following objective function:

$$T = \sum_{j=1}^{n} p_j x_j \qquad (7)$$

The KP also can be extended to consider more restrictions. In example, if cv is the volumetric capacity of the backpack, cz can be added to include its weight capacity. Thus, if w_j represents the volume of the item j, z_j can be used to represent its weight, leading to the following restriction:

$$\sum_{j=1}^{n} z_j x_j \leq cz \qquad (8)$$

Figure 14 presents an overview of the reproduction operator for the GA considered to solve a large KP instance. Note that, due to the binary nature of the decision variable, the crossover and mutation operators can be implemented faster. Then, the configuration settings of the GA are reviewed in **Table 6**.

Based on the instance reported in [24], six random test instances with $N = 250$ items were generated. **Figure 15** presents the mean results for these instances. Error gap assessment was performed with the optimization software Lingo. This led to an error gap of 4.0% which is consistent with the results reported in [24].

4. Genetic algorithm for production scheduling problems

This chapter ends with an application of GA for solving one of the most useful models for manufacturing planning. This model, known as the Permutation Flow-Shop Scheduling Problem (PFSP), consists of finding the optimal sequence of N-jobs to be processed on M-machines [25]. The optimal sequence of jobs is the one that minimizes the make-span of the N-jobs through the M-machines, thus, minimizing the completion time of the last job on the last machine. Note that this sequencing implies two important restrictions: (a) no job can be started on the following machine until it is finished in the previous machine; and (b) a job cannot be started on a machine if it is busy processing another job. As consequence, this is one of the main strategies to reduce idle and waiting times within a workshop [26].

For illustration purposes, **Figure 16** shows an example of a solution for a 5-jobs (a, b, c, d, e) and 3-machines (1, 2, 3) PFSP. Note that each job may take different processing times depending of the assigned machine, and the established sequence remains the same for all machines. Thus, the established sequence has a direct effect on the completion time or *makespan*.

Thus, the information (i.e., processing times) of a PFSP with N-jobs and M-machines is frequently presented as shown in **Table 7**. As in the case of the TSP/CVRP models, the PFSP is also of NP-hard computational complexity, thus, meta-heuristic methods are frequently considered to solve it within reasonable time.

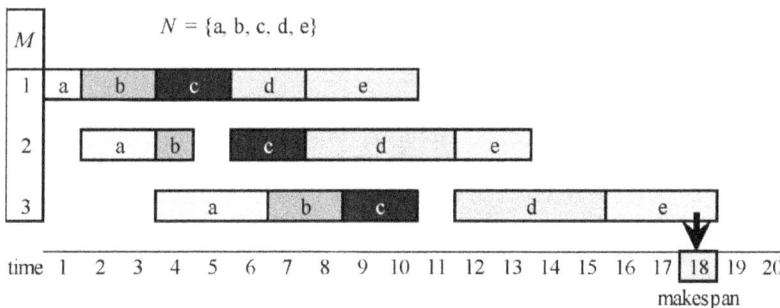

Figure 16.
Example of a 5-jobs, 3-machines PFSP.

Job	Processing Times				
	P_1	P_2	P_3	...	P_M
1	P_{11}	P_{12}	P_{13}	...	P_{1M}
2	P_{21}	P_{22}	P_{23}	...	P_{2M}
...	...	...	...	...	...
N	P_{N1}	P_{N2}	P_{N3}	...	P_{NM}

Table 7.
Data of the PFSP.

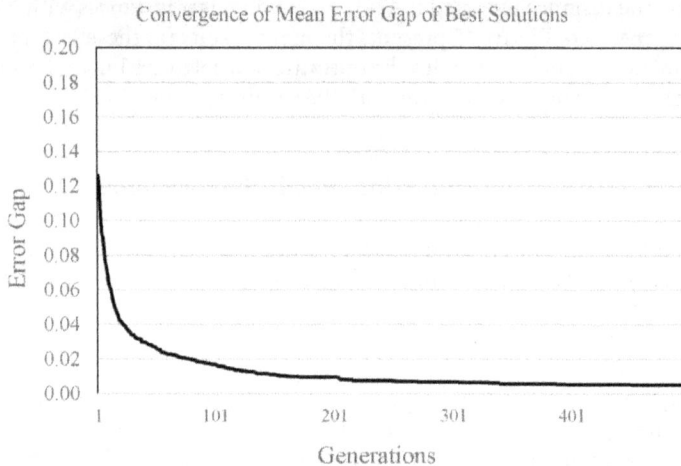

Figure 17.
Mean error gap across 30 randomly selected 20 × 20 PFLP test instances with the GA.

As it is a permutation-based problem, the structure and settings considered for the TSP GA (see **Figure 2** and **Table 1**) were considered for the PFSP with 500 generations. For testing purposes, the library and best results reported in [27] for 30 randomly selected 20-jobs, 20-machines PFSP instances were considered. The results are presented in **Figure 17**.

As observed, the mean error gap reaches the 10% mark at the beginning of the GA, with a final mean error gap of 0.005% by the 500th generation. Thus, the GA can provide near-optimal results for the PFSP.

5. Conclusions and future work

In this chapter the basic elements of a GA were reviewed to describe its application for different logistics and manufacturing problems. The routing problems, beyond the transportation context, can be applied on machine maintenance schemes or material changing services within production plants to minimize operational times. Also, they can be applied to improve the material flow through the warehouse, which is a main facility within the SC. Operations such as order-picking and bin-shelving can be optimized by modeling them as TSP instances [28].

On the other hand, the KP for selection of items is a problem shared with other contexts such as waste reduction in cutting processes, selection of investments and portfolios, decisions for capital budgeting and asset-backed securitization [29]. The

PFSP has been also extended on other fields such as in scheduling of quality control tasks on different machines [30].

Thus, the relevance of solving these combinatorial problems, particularly those of large scale, is very important due to their impact in other science and industrial fields.

Within the search algorithms, the GA can provide very suitable results for these problems. However, as presented in Sections X.2, X.3., and X.4, final performance depends of the type of problem. While the GA can achieve mean error gaps under the 10% mark for TSP/CVRP, for the PFSP the GA can achieve near optimal results under the 1% mark.

These results were supported by extensive experiments which were performed with well-known test databases or libraries. In practice, these experiments also provide important feedback to consider alternative meta-heuristics or develop hybrid approaches for improvement of performance.

This is because, as reviewed, a single meta-heuristic or search algorithm may not be enough to solve all problems if near-optimality is required. In this case, hybridization between different methods have improved on the search mechanisms of meta-heuristics, either deterministic or stochastic. Also, the integration with mathematical programming (which implies an exact solving method) has provided innovative proposals to solve NP-hard problems [31].

Future work is extensive on this field because:

- better solving methods are required due to the presence of increasingly complex combinatorial problems;

- advanced mathematical modeling is required to reduce the complexity of NP-hard problems and thus, make them more suitable to optimization through meta-heuristics or exact methods such as Branch & Bound;

- automatic decision models require the use of Big Data Analysis which, to some extend, depends of meta-heuristic methods.

Thus, as a concluding remark, it can be stated that any advance on these algorithms can impact on different fields. Just to mention an important field within the current industry, meta-heuristics are playing an important role on the implementation of dynamic decision models within Industry/Manufacturing 4.0 systems. Within this context, recent works have reported the application and improvement of these search algorithms for cost-efficient deployment of computing systems in logistics centers [32], dynamic CVRP [33], and development of Digital-Twin platforms [34].

Conflict of interest

The author declares no conflict of interest.

Author details

Gladys Bonilla-Enriquez[1] and Santiago-Omar Caballero-Morales[2*]

1 Technological Institute of Puebla, Puebla, Mexico

2 Autonomous People's University of the State of Puebla, Puebla, Mexico

*Address all correspondence to: santiagoomar.caballero@upaep.mx

IntechOpen

References

[1] Simchi-Levi D, Chen X, Bramel J. The Logic of Logistics Theory, Algorithms, and Applications for Logistics Management. 3rd ed. New Delhi, India: Springer Science+Business Media; 2014. 447 p. DOI: 10.1007/978-1-4614-9149-1

[2] Sivanandam SN, Deepa SN. Introduction to Genetic Algorithms. 1st ed. Berlin: Springer; 2008. 442 p. DOI: 10.1007/978-3-540-73190-0

[3] Singh-Juneja S, Saraswat P, Singh K, Sharma J, Majumdar R, Chowdhary S. TravellingSalesmanProblemOptimization Using Genetic Algorithm. In: Proceedings of the Amity International Conference on Artificial Intelligence (AICAI 2019); 4-6 February 2019; Dubai, United Arab Emirates, United Arab Emirates: IEEE; 2019. p. 264-268

[4] Crama Y, van de Klundert J, Spieksma FCR. Production planning problems in printed circuit board assembly. Discrete Appl Math. 2002;**123**:339-361

[5] Bertsimas DJ, Simchi-Levi D. A new generation of vehicle routing research: robust algorithms, addressing uncertainty. Oper Res. 1996;**44**(2):286-304

[6] Tsung-Sheng C, Yat-Wah W, Wei TO. A stochastic dynamic travelling salesman problem with hard time windows. Eur J Oper Res. 2009;**198**:749-759

[7] Bharati TP, Kalshetty YR. A hybrid method to solve travelling salesman problem. IJIRCCE. 2016;**4**(8):15148-15152

[8] Halim AH, Ismail I. Combinatorial Optimization: Comparison of Heuristic Algorithms in Travelling Salesman Problem. Arch Computat Methods Eng. 2019;**26**:367-380. DOI: 10.1007/s11831-017-9247-y

[9] Reinelt G. TSPLIB 95 [Internet]. 1995. Available from: http://comopt.ifi.uni-heidelberg.de/software/TSPLIB95/ [Accessed: 2021-01-20]

[10] Nguyen HD, Yoshihara I, Yamamori K, Yasunaga M. Implementation of an Effective Hybrid GA for Large-Scale Traveling Salesman Problems. IEEE TRANSACTIONS ON SYSTEMS, MAN, AND CYBERNETICS—PART B: CYBERNETICS. 2007;**37**(1):92-99. DOI: 10.1109/TSMCB.2006.880136

[11] Toth P, Vigol D. Models, relaxations and exact approaches for the capacitated vehicle routing problem. Discrete Applied Mathematics. 2002;**123**(1-3):487-512. DOI: 10.1016/S0166-218X(01)00351-1

[12] Zhang DZ, Lee CKM. An Improved Artificial Bee Colony Algorithm for the Capacitated Vehicle Routing Problem. In: Proceedings of the 2015 IEEE International Conference on Systems, Man, and Cybernetics; 9-12 October 2015; Kowloon, China: IEEE; 2015. p. 2124-2128. DOI: 10.1109/SMC.2015.371

[13] Kwon YJ, Kim JG, Seo J, Lee DH, Kim DS. A Tabu Search Algorithm using the Voronoi Diagram for the Capacitated Vehicle Routing Problem. In: Proceedings of the 5th International Conference on Computational Science and Applications (ICCSA 2007); 26-29 August; Kuala Lampur, Malaysia: IEEE; 2007. p. 480-485. DOI: 10.1109/ICCSA.2007.11

[14] Jin J, Crainic TG, Lokketangen A. A parallel multi-neighborhood cooperative tabu search for capacitated vehicle routing problems. European Journal of Operational Research. 2012;**222**(3):441-451. DOI: 10.1016/j.ejor.2012.05.025

[15] Nazif H, Lee L. Optimised crossover genetic algorithm for capacitated vehicle

routing problem. Applied Mathematical Modelling. 2012;**36**:2110-2117. DOI: 10.1016/j.apm.2011.08.010

[16] Mari F, Mahmudy WF, Santoso PB. An Improved Simulated Annealing for the Capacitated Vehicle Routing Problem (CVRP). Jurnal Ilmiah Kursor. 2018;**9**(3):119-128. DOI: 10.28961/kursor.v9i3.178

[17] Ilhan I. A population based simulated annealing algorithm for capacitated vehicle routing problem. Turkish Journal of Electrical Engineering & Computer Sciences. 2020;**28**:1217-1235

[18] Khaddar-Bakhayt AG, Al-Sattar HA, Abbas IT. Solving CVRP by Using Two-stage (DPSOTS) Algorithm. Global Journal of Pure and Applied Mathematics. 2017;**13**(6):1553-1567

[19] Uchoa E, Pecin D, Pessoa A, Poggi M, Vidal P, Subramanian A. New Benchmark Instances for the Capacitated Vehicle Routing Problem. European Journal of Operational Research. 2017;**257**(3):845-858. DOI: 10.1016/j.ejor.2016.08.012

[20] Uchoa E, Pecin D, Pessoa A, Poggi, M, Subramanian A, Vidal T. CVRPLIB [Internet]. 2014. Available from: http://vrp.atd-lab.inf.puc-rio.br/index.php/en/ [Accessed: 2021-01-20]

[21] Baker BM, Ayechew MA. A genetic algorithm for the vehicle routing problem. Computers & Operations Research. 2003;**30**(5):787-800. DOI: 10.1016/S0305-0548(02)00051-5

[22] Sanchez-Sierra ST, Caballero-Morales SO, Sanchez-Partida D, Martinez-Flores JL. Facility Location Model with Inventory Transportation and Management Costs. Acta Logistica. 2018;**5**(3):79-86

[23] Martello S, Toth P. Knapsack Problems: Algorithms and Computer Implementations. 1st ed. England: John Wiley & Sons Ltd; 1990. 308 p

[24] Bonilla-Enriquez G, Sanchez-Partida D, Caballero-Morales SO. Algoritmo genetico para el problema logistico de asignacion de la mochila (Knapsack Problem). Research in Computing Science. 2017;**137**:157-168

[25] Singhal E, Singh S, Dayma A. An Improved Heuristic for Permutation Flow Shop Scheduling (NEH ALGORITHM). International Journal of Computational Engineering Research. 2012;**2**(6):30-36

[26] Rosas-Gonzalez A, Clemente-Guerrero DM, Caballero-Morales SO, Flores-Juan JC. Machines Permutation Flow-Shop Scheduling Problem with Break-Down Times. International Journal of Computer Applications. 2013;**83**(1):1-6. DOI: 10.5120/14409-2488

[27] Watson JP, Barbulescu L, Whitley DL, Howe AE. Contrasting structured and random permutation flow-shop scheduling problems: Search space topology and algorithm performance. INFORMS Journal on Computing. 2002;**14**(2):98-123

[28] Bartholdi JJ, Hackman ST. Warehouse & Distribution Science. Release 0.96. Atlanta, Georgia, USA: The Supply Chain and Logistics Institute; 2014. 323 p

[29] Kellerer H, Pferschy U, Pisinger D. Knapsack Problems. 1st ed. Germany: Springer-Verlag Berlin Heidelberg; 2004. 548 p. DOI: 10.1007/978-3-540-24777-7

[30] Erseven G, Akgün G, Karakaş A, Yarıkcan G, Yücel Ö, Öner A. An Application of Permutation Flowshop Scheduling Problem in Quality Control Processes. In: Proceedings of the International Symposium for Production Research (ISPR 2018); 28-31 August; Vienna, Austria: Springer; 2018. p. 849-860. DOI: 10.1007/978-3-319-92267-6_68

[31] Boschetti MA, Maniezzo V, Roffilli M, Röhler AB. Matheuristics: Optimization, Simulation and Control. In: Proceedings of the International Workshop on Hybrid Metaheuristics (HM 2009); 16-18 January; Concepcion, Chile: Springer Verlag; 2009. p. 171-177. DOI: 10.1007/978-3-642-04918-7_13

[32] Li CC, Yang JW. Cost-Efficient Deployment of Fog Computing Systems at Logistics Centers in Industry 4.0. IEEE Transactions on Industrial Informatics. 2018;**14**(10):4603-4611. DOI: 10.1109/TII.2018.2827920

[33] Abdirad M, Krishnan K, Gupta D. A two-stage metaheuristic algorithm for the dynamic vehicle routing problem in Industry 4.0 approach. Journal of Management Analytics. 2020. DOI: 10.1080/23270012.2020.1811166

[34] Balderas D, Ortiz A, Mendez E, Ponce P, Molina A. Empowering Digital Twin for Industry 4.0 using metaheuristic optimization algorithms: case study PCB drilling optimization. The International Journal of Advanced Manufacturing Technology. 2021. DOI: 10.1007/s00170-021-06649-8

Chapter 3

An Adaptive Task Scheduling in Fog Computing

Dinesh G. Harkut, Prachi Thakar and Lovely Mutneja

Abstract

Internet applications generate massive amount of data. For processing the data, it is transmitted to cloud. Time-sensitive applications require faster access. However, the limitation with the cloud is the connectivity with the end devices. Fog was developed by Cisco to overcome this limitation. Fog has better connectivity with the end devices, with some limitations. Fog works as intermediate layer between the end devices and the cloud. When providing the quality of service to end users, scheduling plays an important role. Scheduling a task based on the end users requirement is a tedious thing. In this paper, we proposed a cloud-fog task scheduling model, which provides quality of service to end devices with proper security.

Keywords: ANN, fuzzy logic, fog computing, IoT, QoS, K-means clustering

1. Introduction

Cloud computing is very popular in the technology world as it provides numerous useful services to end users. Cloud computing is based heavily on virtualization technology. Cloud computing provides many features such as huge processing power, great storage provision, and pay-per-use model. Cloud computing has many desirable features such as flexibility, scalability, performance-cost efficiency, and ease of test, adopting and deploying new technologies [1].

In spite of all these services, there are some drawbacks of cloud computing that cannot be ignored. For examples, the cloud and users are physically far away from each other that induce intolerably delay, again there can be a shortage of resources for executing the tasks, many resources could remain idle even though tasks need to be processed, etc. [1].

Internet of Things (IoT) is an emerging technology. It requires latency-aware computation for real-time application processing. In IoT environments, devices connected to it generate a huge amount of data, which are generally referred to as big data. IoT devices generated data are generally processed in a cloud infrastructure because of the on-demand services and scalability features of the cloud computing. However, processing IoT application requests on the cloud is not an efficient solution for some IoT applications, especially time-sensitive ones. To address this issue, Fog computing, which is a middle layer between cloud and IoT devices, was proposed. In Fog computing environment, IoT devices are connected to Fog devices. These Fog devices are located in close proximity as compared to cloud to users and are responsible for intermediate computation and storage [2].

There are many challenges when we are working in fog computing environment. One of the challenges is task scheduling. Tasks are broadly classified into two category, dependent task and independent task. While performing task scheduling in fog, the category of tasks plays a vital role.

Task scheduling depends on the many criteria based on user's requirements. For example, healthcare-related task. In such type of task time is a vital factor. Delay in such type of task is not acceptable, so to manage such type of tasks, many task scheduling algorithms have been proposed. Task scheduling involves scheduling of resources, such as CPU, memory. Depending on the type of task, algorithm may varies. The basic idea behind task scheduling is to give the user QoS (Quality of Service).

2. Literature review on task scheduling in fog-cloud environment

Author in [3] have used Q learning algorithm in cloud computing for allocating the task to the virtual machine. In this paper, we have compared their algorithm with FIFO, greedy, random, mix algorithm. The proposed model is divided into three parts: tasks transmission, task allocation, and task execution.

Resource Management and task scheduling are very important tasks in cloud. The traditional scheduling algorithm has low resource utilization and more response. Rather than using single scheduling algorithm, multiple scheduling algorithms are used. The selection of one of the scheduling algorithm is done using machine learning classification. Six scheduling algorithms are considered here. FCFS, priority scheduling conservative migration supported backfilling, aggressive migration supported backfilling, and priority-based consolidation. Selection of particular algorithm based on environment and task is done using machine learning classification [4].

Two reinforcement schedulings were introduced for resource scheduling, online resource scheduling deepRM2 and offline resource scheduling DeepRM_off. Then the comparison of these two algorithms has done with the DeepRM and the heuristic algorithm. Two resources are considered in this CPU and memory. Image is given as input for training process [5].

Three approaches for tasks scheduling are discussed and compared in this paper. PSO algorithm, genetic algorithm, modified PSO algorithm. Modified PSO algorithm is nothing but the old PSO with the merging of SJFP for generating initial population in order to minimize makespan. The result shows that the modified PSO outperforms the other two algorithms [6].

Author in [7] suggested a new technique to schedule the Jobs or tasks in Big Data cluster. The uniqueness of this proposed method is that it basically focuses on the resources utilization and the type of Scheduled job altogether. The clusters used for experimentation of the proposed method are homogeneous. The given algorithm assigns task to the data node based on the type of job and based on the data node resource load.

K-means clustering algorithm is used for grouping the virtual machine and task [8]. The categorization of virtual environment is done on the basis of available application in each machine. Four parameters are considered for task selection, task length, user priority, deadline, and cost. K-means clustering technique is used for virtual machine as well as for task.

To select a proper task scheduling algorithm for better performance in cloud computing is a critical task. Author in [9] suggested a Framework for the above problem. Author suggested that the decision of which task scheduling algorithm is suitable for a particular task should be taken by machine learning algorithm.

The Basic concept in [10] is distribution of task to different fog nodes. The proposed approach performance is compared with PSO and GA. The proposed approach divides the complete task into two part reproduction behavior, food source foraging behavior. The implementation of BLA is done in C ++. The proposed algorithm outperforms in terms of CPU execution time, allocation of memory, and therefore, the cost function. The limitation of this approach was it does not give any solution on dynamic job scheduling, and again here they consider stationary fog servers.

Author in [11] proposed test and selection technique to select the best algorithm for scheduling. The hyper-heuristic algorithm is divided into two phases, training phase and testing phase. The basic objective is to find out the best algorithm for workflow scheduling. The author [11] considered four algorithms for the purpose of selection, genetic algorithm, particle Swarm Optimization, ant Colony Optimization, and annealing algorithm. FogSim is used for fog computing environment, and cloud SIM is used for cloud environment.

Author in [12] suggested Ant colony algorithm for scheduling. The tasks are grouped according to two criteria, minimum cost and minimum end time service. Also the prioritization of task is done based on the above two criteria. The ant Colony algorithm is used to select optimal virtual machine for executing the task.

The main focus is on multi-resource fairness in task and to achieve ultralow task latency for fog computing system. Author in [13] proposed fair TS online test scheduling model. Author uses DRL technique to gain experience and based on that the Fair TS model is developed. Researchers [13] claim that their model balances the time and resources. The main challenge of this paper is to perform online task scheduling. The number of tasks is already fixed. For multi-resource fairness in fog computing system dominant resource fairness policies are adopted.

Different fog node has different processing abilities, for example, strong fog node with considerable resources can solve the complex problem easily. But such type of scenario development is a problem in task scheduling. This problem is addressed in this paper. A new task scheduling strategy is suggested in this paper. Hybrid heuristic algorithm is proposed for tasks scheduling. The hybrid heuristic algorithm is combination of two algorithms, improved particle Swarm Optimization and improved ant Colony Optimization [14].

Issues related to mobile crowd sensing task in fog computing are discussed. A deep reinforcement scheduling solution is provided to solve this problem. It is a self-adaptive model. Three-layer hierarchical structure of fog computing is discussed. To solve task scheduling problem in fog computing, a task scheduler is added in the cloud layer to decide the scheduling strategy for fog computing [15].

Three-layered structure is introduced: terminal layer, which consists of mobile devices; fog layer, which consists of task scheduling cluster and resource integration model; core layer composed of cloud resource provider. Scheduling is done in the middle layer. A new scheduling method was introduced "I-FASC" to determine the characteristics of task and resources. An improved genetic algorithm is proposed, which is an improvement over the firework algorithm, which introduced the explosion radius detection mechanism of Fireworks to avoid disappearance of optimal solution [16].

The problem with delay-sensitive application such as smart health required to transfer large amount of data to cloud, so it reduces the performance. To resolve that, fog computing is introduced. But in fog, there should be some mechanism to manage the task and resource as well as security. To achieve this, a cost-aware genetic-based task scheduling algorithm is proposed [17].

Two characteristics of Intelligence are considered, device-driven and human-driven in IOT-based computing scenario. For demonstration purposes, two cases are considered. The first case machine learning algorithm is used to study the human behavior based on that scheduling is done that is identifying the priority of the task whether the task is important or not, and if it is important in that case, give the resource to that task. In second case, an algorithm is designed for the end user device to select the offloading decision, that is, to identify whether to process the task or discard it to minimize energy consumption of fog node and to minimize the latency [18].

In the three-tier architecture, the end devices are at the lowest level, fog is the intermediate layer, and the top most layers consist of cloud. Intelligent virtual machine is created by using Bayesian method to classify task. The FBCS algorithm outperforms when compared with the FCFS and delay priority algorithm. Two algorithms are designed: first for task classification and second for updating processing power of the device [19].

When we are talking about scheduling in cloud computing that basically means we are focusing on how to improve the use of resources and reduce the time to complete a job. The cost to do certain job depends on time and exchange of data. To reduce the cost of the user, decrease the volume of data sent to the cloud. This was the main idea behind the creation of fog. The IoT devices can connect to the cloud through fog nodes [20] .

The tasks are scheduled based on lower delay. In this paper, the problem related to task scheduling in fog computing is discussed. The dynamic scenario resulted from user mobility brings a dynamic computing demand at edge devices. The scheduling strategies should be designed based on the different application classes according to demand coming from the mobile user [21].

A metaheuristic algorithm based on a Harris Hawk Optimization based on a local search strategy for task scheduling in fog computing is proposed to improve the quality of service provided to the user in industrial IoT application [22].

For scheduling purpose, an optimized knapsack algorithm is proposed, which is based on symbiotic organism search algorithm [23].

An improved apriori algorithm "I-Apriori" is proposed based on apriori algorithm. A novel task scheduling TSFC algorithm is proposed. The association rules are generated by the I-apriori algorithm. The TSFC algorithm is based on I-apriori algorithm [24].

Tasks scheduling problem is discussed to reduce the cost of Edge computing sys-tem. The focus of this paper is to minimize the cost while satisfying the delay requirement of the entire task. For that a two-star scheduling cost Optimization algorithm is proposed (TTSCO) [25].

The focus of this paper is on how to reduce the power consumption in edge computing while meeting the resource and delay constraints [26].

Task scheduling algorithm based on delay model is suggested. Others claim that the delay model based on little's law is in accurate. So the authors suggest a delay model without using little's law. Then a life Lyapunov function of delay is defined based on that a task scheduling algorithm is proposed to minimize the delay [27].

The focus [28] is to provide the quality of service to the user and to improve the performance of scheduling. An application-aware scheduling algorithm is proposed.

A scheduling algorithm for cloudlet for utilization of the available resources is suggested. The proposed algorithm is based on ant Colony Optimizations algorithm [29].

Quality of service was the main motive behind a grouped tasks scheduling algorithm. The GTS algorithm divides the task into categories. User task, task latency, task size, task type are the parameters used for categorizing a task. GST first chooses which category to be executed and then chooses the task with mini-mum time to be executed in the category [30].

A new scheduling algorithm is proposed, FCAP. This new algorithm is combination of two algorithms: Fuzzy C-means clustering algorithm and PSO particle Swarm Optimization algorithm. FCAP is used to cluster the resources. The main idea behind this algorithm is to provide quality of service to the users [31].

A reinforcement learning agent is proposed that horizontally scales container's instances based on the demand of user and available fog resources. FScalar is integrated in kubernetes cluster architecture. Also the use of SARSA to build a scalar agent is proposed [32].

Author in [2] studied the current trend of fog computing as well as the architecture of fog computing. Author also explained the limitation of such architecture and pointed out the deployment issue of services in fog. Efficiently placing a new service without affecting the running one is the biggest problem with the fog architecture.

RSU (roadside unit) acts as an immobile fog node. The responsibility of RSU is request processing and decision-making for task scheduling. Author in [33] investigated the tasks scheduling and resource allocation from the viewpoint of service-oriented architecture (SOA). Tasks scheduling is based on scheduling chain.

A novel energy efficient fog computing Framework is proposed by the author. The homogeneous fog network is considered for framework. The main focus of the paper is on Energy Efficiency for task scheduling. Author in [34] Suggested maximum energy-efficient task scheduling algorithm MEETS in homogeneous fog network.

Parallel execution of tasks in heterogeneous fog network is suggested. New concept PE processing efficiency is defined, which includes computing resources and communication capabilities. DATS algorithm is introduced to minimize the service delay in heterogeneous fog network. The two key components of DATS are PCRC (progressive computing resource competition) to obtain stable resource allocation result and second is STS (synchronize task scheduling) [35].

An adaptive multi-objective Optimization testing task scheduling method for fog computing is proposed. The two objectives of these proposed algorithms are task scheduling and resource scheduling with minimum task execution time and resource cost [36].

A new concept is introduced [37], "region". Region is nothing but the collection of fog node. Basically the fog nodes are divided into region based on the requirement of the user. A task scheduling algorithm for region-based cloud (FBRC) is proposed [37].

A best selection of fog device for offloading the task by considering the time and energy consumption is a very serious challenge. To address this problem, a module placement method by classification and regression tree algorithm is proposed. The parameters for selecting the best fog node for the task are authenticity, integrity, confidentiality, speed, cost, capacity, and availability. Model placement is based on Markov chain process [38].

A tool kit that can automatically simulate the complex network topology and different type of computing resources as well as automatically execute user submitted workflow application and compare the performance of different computation offloading and task scheduling strategy for workflow is suggested [39].

A Ranking-based task scheduling algorithm using linguistic and fuzzy quantified in fog cloud network preposition is proposed. This algorithm is compared with distance-based algorithm, price-based algorithm, and latency-based algorithm [40].

Load balancing in cloud and fog is suggested in this paper. Cuckoo search by using levy walk distribution and flower pollination is proposed for load balancing. The motto is to reduce the delay and to overcome the latency issue [41].

The task is assigned with the priority depend on the deadline of the task. Preemption of the task is not possible after assigning it to the particular fog node [42].

Literature review

Paper no	Title	Year	Basic concept	Evaluation parameter					Independent/ dependent
				Bandwidth	Cost	Energy consumption	Time	Throughput/ latency	
10	Fog computing job scheduling optimization based on bees Swarm	2018	The basic concept in this paper is distribution of task to different fog node. The proposed approach divide the complete task into 2 part. Reproduction behavior, food source foraging behavior		Cost		Time		Not mentioned
11	A Hyper Heuristic Algorithm for scheduling of fog networks	2017	In this paper a test and select technique is used to select best algorithm for scheduling	Bandwidth	Cost	Energy consumption		Throughput	Not mentioned
12	Providing A ne scheduling method theme fog network using the ant colony algorithm	2019	The tasks are grouped according to 2 criteria, minimum cost and minimum end time service		Cost		Time	Throughput	Not mentioned
13	Online scheduling for fog computing with multi resource fairness	2019	Deep reinforcement technique is used					Throughput	Not mentioned
14	Task scheduling based on hybrid heuristic algorithm for smart production line with fog computing	2019	Two algorithm are combined, improved PSO and improved ACO			Energy consumption	Time		Not mentioned
15	Deep reinforcement scheduling for mobile crowd sensing in fog computing	2019	Three layered hierarchical structure of fog computing is discussed. A scheduler is added for scheduling decision in the first layer.	Bandwidth					Not mentioned

Literature review

Paper no	Title	Year	Basic concept	Evaluation parameter						Independent/ dependent
				Bandwidth	Cost	Energy consumption	Time	Throughput/ latency		
16	Task scheduling algorithm based on improved firework algorithm in fog computing	2020	IFASC algorithm is proposed to determine the characteristics of task & resources. [IFA] improved genetic algorithm is proposed				Time			Not mentioned
17	Cost aware task scheduling in fog-cloud environment	2020	A cost aware genetic based task scheduling algorithm is proposed				Time			Not mentioned
18	Enabling intelligence in fog computing to achieve energy and latency reduction	2019	Two characteristics of intelligence is consider, device-driven and human-driven			Energy consumption		Latency		Not mentioned
19	Energy saving scheduling in fog based iot application by Bayesian approach	2019	Two algorithm design 1. For task classification. 2. For updating processing power of device		Cost	Energy consumption				Not mentioned
20	Smart fog: fog computing Framework for unsupervised clustering Analytics in wearable internet of things	2017	Decrease the volume of send data to cloud	Bandwidth			Time			Not mentioned
21	Mobility aware application scheduling in fog computing	2017	Dynamic scheduling							Not mentioned

Literature review

Paper no	Title	Year	Basic concept	Evaluation parameter					Independent/ dependent
				Bandwidth	Cost	Energy consumption	Time	Throughput/ latency	
22	Energy-aware Marine Predator algorithm for task scheduling in IoT based fog computing application	2020	A meta heuristic algorithm based on Harish hawks Optimization based on local search strategy for task scheduling in fog computing is proposed		Cost	Energy consumption	Time		Independent
23	Scheduling of fog network with optimized knapsack by symbiotic organism search	2017	A new KnapSOS algorithm is proposed	Bandwidth	Cost	Energy consumption	Time		Not mentioned
24	A task scheduling algorithm based on classification mining in fog computing environment	2018	Improved apriori algorithm is proposed to generate Association rules.				Time		Not mentioned
25	Cost efficient scheduling for delay sensitive task in edge computing system	2018	Minimize the cost while satisfying the delay requirements of all task		Cost		Time		Not mentioned
26	A scheduling strategy for reduce power consumption in Mobile Edge computing	2020	Edge nodes are divided into master and slave nodes			Energy consumption			Not mentioned
27	A more accurate delay model based task scheduling in cellular edge computing systems	2019	Delay model without using little's Law				Time		Not mentioned
28	Application-aware task scheduling in heterogeneous edge computing	2019	One master node and multiple slave node				Time	Latency	Not mentioned

Literature review

Paper no	Title	Year	Basic concept	Evaluation parameter						Independent/ dependent
				Bandwidth	Cost	Energy consumption	Time	Throughput/ latency		
29	Churn-resilient task scheduling in a tired IOT infrastructure	2019	The proposed algorithm is based on ant colony optimization to tackle the DYNAMICS of service provider							Not mentioned
30	Grouped task scheduling algorithm based on quality of service in cloud computing network	2016	The task are divided into categories				Time	Latency		Not mentioned
31	Methods of resource scheduling based on fuzzy clustering in fog computing	2019	The proposed algorithm is combination of Fuzzy c means clustering algorithm and PSO algorithm	Bandwidth	Cost		time	Latency		Not mentioned
32	FScalar: automatic resource scaling of container in for cluster using reinforcement learning	2020	Reinforcement learning agent is proposed that horizontally scales containers instances based on the demand of user unavailable fog resources.							Not mentioned
2	Fog computing: survey of trends architecture and Research direction	2016	Study the current trend of fog computing as well as the architecture of fog computing							Not mentioned
33	RSU- empowered resource pooling for task scheduling in vehicular fog computing	2020	Task scheduling is based on scheduling chain	Bandwidth	Cost		Time			Not mentioned
34	MEETS: maximum energy efficient as scheduling in homogeneous fog network	2018	Maximize energy efficiency for task scheduling			Energy consumption				Not mentioned

Literature review

Paper no	Title	Year	Basic concept	Evaluation parameter					Independent/ dependent
				Bandwidth	Cost	Energy consumption	Time	Throughput/ latency	
35	DATS: dispersive stable task scheduling in heterogeneous fog network	2018	the two key component of DATS are PCR progressive computing resource competition to obtain stable resource allocation result and second is sts synchronize task scheduling				Time		Independent
36	A multi-objective task scheduling method for fog computing in cyber physical social service	2020	Task scheduling and resource scheduling with minimum task execution time and resource cost		Cost		Time		Not mentioned
37	FBRC: optimization of task scheduling in fog based region and cloud	2017	The fog note are divided into regions	Bandwidth				Latency	Not mentioned
38	Task offloading in mobile fog computing by classification and regression tree	2019	A best selection of fog device for offloading the task by considering the time and energy consumption			Energy consumption	Time	Latency	Not mentioned
39	Fog workflows: an automated simulation toolkit for workflow performance evaluation in fog computing	2019	Automatically simulate the complex network topology and different types of computing resources as well as automatically execute user submitted workflow application			Energy consumption		Latency	Not mentioned
41	Cloud and fog based integrated environment for load balancing	2019	Cuckoo search by using levy walk distribution and Flower pollution is proposed for load balancing					Latency	Not mentioned
42	An optimal task scheduling toward minimized cost and response time in a fog computing infrastructure	2019	The task is assigned with the priority depend on the deadline of the task .		Cost			Throughput	Independent

Literature review

Paper no	Title	Year	Basic concept	Evaluation parameter					Independent/ dependent
				Bandwidth	Cost	Energy consumption	Time	Throughput/ latency	
43	Parallel scheduling of multiple tasks in heterogeneous fog network	2019	For scheduling the task a distributed task scheduling algorithm was developed via gauss seidel type method				Time	Latency	Not mentioned
44	Online task scheduling and resource allocation for intelligent NOMA-based industrial internet of things	2020	A non-orthogonal multiple access based fog computing Framework for industrial IoT system is proposed			Energy consumption	Time		Not mentioned
45	Task scheduling and resource allocation in fog computing based on container for smart manufacturing	2018	A container based task scheduling algorithm for delay sensitive and high concurrency characteristics of fog computing is proposed					Latency	Not mentioned
46	Deadline-aware fair scheduling for offloaded tasks in fog computing with inter fog dependency	2019	Task with the different deadline are considered. 2 queues are consider. For scheduling in the queue Lyapunov drift plus penalty function is used.					Latency	Not mentioned
47	A method based on combination of lexity and ant colony system for cloud for task scheduling	2019	Laxity based priority algorithm is used for deciding priority of the task. To minimize energy consumption ant colony method is proposed			Energy consumption	Time		Dependent
48	Security aware scheduling in fog computing by hyper-heuristic algorithm	2017	Focuses on workflow scheduling problem			Energy consumption	Time		Not mentioned
49	Delay minimized task scheduling in fog enabled IoT networks	2020	Scheduling of delay sensitive task				Time		Not mentioned

Literature review

Paper no	Title	Year	Basic concept	Evaluation parameter					Independent/ dependent
				Bandwidth	Cost	Energy consumption	Time	Throughput/ latency	
50	A novel energy aware scheduling and load balancing technique based on fog computing	2019	4 criteria are considered in the proposed algorithm.			Energy consumption	Time		Not mentioned
51	Neural task scheduling with reinforcement learning for fog computing system	2019	Deep reinforcement learning and pointer network architecture are combined to propose neural task scheduling					Throughput	Not mentioned
3	Reinforcement learning based foresighted task scheduling in cloud	2018	The proposed model is divided into three,; part task transmission task allocation task execution				Time	Throughput	Not mentioned
4	Task scheduling in cloud using machine learning classification	2015	The selection of scheduling algorithm is done using machine learning classification						Not mentioned
5	A new approach for resource scheduling with deep reinforcement learning	2018	Two reinforcement scheduling was introduced for resource scheduling. Online resource scheduling offline resource scheduling.				Time	Latency	Not mentioned
6	Task scheduling using modified PSO algorithm in cloud computing environment	2017	Three approaches for task scheduling is discussed and compared, PSO algorithm genetic algorithm modified PSO				Time		Not mentioned
7	A new approach for scheduling task and /or job in Big Data cluster	2019	The given algorithm assign task to the data node based on the type of job and based on the data node resource load						Not mentioned

Literature review

Paper no	Title	Year	Basic concept	Evaluation parameter					Independent/ dependent
				Bandwidth	Cost	Energy consumption	Time	Throughput/ latency	
8	A credits based scheduling algorithm with K-means clustering	2018	K-means clustering algorithm is used for grouping the virtual machine and task				Time		Not mentioned
9	Framework for task scheduling in cloud using machine learning technique	2020	How to select proper task scheduling algorithm based on type of task is depend on the ML						Not mentioned

Table 1.
Literature summary.

Tasks are divided into subtask, and to manage the subtask is challenging issue. To handle this challenge, a generalized Nash equilibrium game called parallel scheduling of multiple tasks is developed. For scheduling the task, a distributed task scheduling algorithm was developed via Gauss Spidel-type method [43].

Non-orthogonal multiple accesses-based fog computing framework for industrial IoT system is proposed. Here the task offloading is based on NOMA to the helper node to minimize the delay and energy consumption [44].

A container-based task scheduling algorithm for delay-sensitive and high-concurrency characteristic of fog computing is proposed. The tasks execution is divided into two steps: first to determine whether to accept or reject; second if accepted, then where to forward the task on fog node or cloud. For resource real-location, a reallocation mechanism is proposed [45].

Tasks with different deadlines are considered. The main objective is to minimize failure probability to meet the different delay deadline. Two queues are considered, low and high-priority queues. For scheduling in the queues, Lyapunov drift plus penalty function is used [46].

To handle the sensitivity of task delay, the laxity-based priority algorithm is suggested. This algorithm is used to decide the priority of the task based on the deadline. Again to minimize energy consumption, an Optimization algorithm based on ant Colony is proposed [47].

The proposed method is based on HH algorithm; it generally focuses on work-flow scheduling. The proposed algorithm shows that it reduces the energy consumption and execution time of the task [48].

Delay-sensitive task is considered. DMTO is proposed to identify the optimal subtask size and the TN transmission power [49].

Four criteria are considered in the proposed algorithm: energy dynamic, threshold, waiting time of the task, and communication delay. These criteria are divided into two groups, and based on that, two scheduling and load balancing procedures are performed [50].

Online task scheduling problem in fog computing is discussed. The main focus is to minimize the task slowdown. Deep reinforcement learning and pointer network architecture are combined to propose neural task scheduling [51].

Author in [1] basically focuses on how to reduce the cost. The proposed algorithm efficiently prioritizes the task according to their delay or tolerance level result in higher throughput, which leads to reduce in overall response time and cost (**Table 1**).

3. Motivation

As we already know that fog is a middle layer between cloud and user. The user's requirement is always QoS. QoS depends on the parameters such as bandwidth, energy consumption, latency, throughput, and cost. So basically fog has to fulfill these requirements of users. Again Fog has limitation such as limited resources and capabilities, but it has an advantage of being nearer to the end devices, which makes it powerful in many aspects such as less latency, less power consumption, and proper utilization of bandwidth. Decision-making on task scheduling is a trending research area. How accurately you can predict the best algorithm on the basis of user's requirement is a challenging issue in fog. Machine learning is making very much progress in this domain. This thing motivates us to use Machine learning algorithm for task scheduling in cloud.

4. Proposed work

The Decision of selecting best algorithm based on the requirement is complex work. For taking the decision on which task to schedule first is completely dependent on the type of task. Again identifying the type of task is another challenge. The first part in task scheduling is to identify the type of task, and then we can perform the actual task scheduling. Task scheduling in fog is mandatory because the end user requires the Quality of Service. The parameters that are considered for QoS are bandwidth, latency, robustness, time, cost, and energy consumption.

Computational Intelligence (CI) is a sub-branch of AI. CI can be considered as the study of adaptive mechanisms to enable or facilitate intelligent behavior in complex and changing environments. Computational Intelligence techniques include fuzzy sets, ANN, Evolutionary computing, swarm intelligence, and artificial immune system. CI is a set of nature-inspired computational methodologies and approaches to address complex real-world problems. The powerful feature of CI is its adaptive nature.

5. Conclusions

In this paper, we have reviewed different existing models and techniques for task scheduling in cloud-fog environment. In first half of the paper, we discussed the limitations and advantages of fog. In second half of the paper, we reviewed the existing technique for task scheduling in fog. By analyzing the existing system, we proposed a CI-based task scheduling model in fog, which will adapt to varying requirements of QoS dynamically.

Author details

Dinesh G. Harkut*, Prachi Thakar and Lovely Mutneja
Department of Computer Science and Engineering, Prof Ram Meghe College of
Engineering and Management, India

*Address all correspondence to: dg.harkut@gmail.com

References

[1] Choudhari T, Moh M, Moh T-S. Prioritized task scheduling in fog computing. In: Proceedings of the ACMSE 2018 Conference, New York. 2018. pp. 1-8. DOI: 10.1145/3190645.3190699

[2] Naha RK et al. Fog computing: Survey of trends, architectures, requirements, and research directions. IEEE Access. 2018;**6**:47980-48009. DOI: 10.1109/ACCESS.2018.2866491

[3] Mostafavi S, Ahmadi F, Sarram M. Reinforcement-Learning-Based Fore-Sighted Task Scheduling in Cloud Computing. 2018. Available from: http://bit.ly/3X7cWG5

[4] Pathak GR. Task Scheduling in the Cloud Using Machine Learning Classification. 2015. Available from: http://bit.ly/3EGGgMB

[5] Ye Y. et al. A new approach for resource scheduling with deep reinforcement learning. 2018. [Accessed: September 6, 2020]

[6] Abdi S, Motamedi S, Sharifian S. Task scheduling using modified PSO algorithm in cloud computing environment. In: Int Conf Mach Learn Electr Mech Eng. 2014. pp. 37-41. Available from: http://bit.ly/3OhNZnw

[7] Hadjar K, Jedidi A. A new approach for scheduling tasks and/or jobs in big data cluster. In: 4th MEC International Conference on Big Data and Smart City (ICBDSC). 2019. pp. 1-4. DOI: 10.1109/ICBDSC.2019.8645613

[8] Sharma V, Bala M. A credits based scheduling algorithm with K-means clustering. In: 2018 First International Conference on Secure Cyber Computing and Communication (ICSCCC). 2018. pp. 82-86. DOI: 10.1109/ICSCCC.2018.8703201

[9] Shetty C, Sarojadevi H. Framework for task scheduling in cloud using machine learning techniques. In: 2020 Fourth International Conference on Inventive Systems and Control (ICISC). 2020. pp. 727-731. DOI: 10.1109/ICISC47916.2020.9171141

[10] Bitam S, Zeadally S, Mellouk A. Fog computing job scheduling optimization based on Bees Swarm. Enterpreneurship. 2017;**12**:1-25. DOI: 10.1080/17517575.2017.1304579

[11] Kabirzadeh S, Rahbari D, Nickray M. A hyper heuristic algorithm for scheduling of fog networks. In: 2017 21st Conference of Open Innovations Association (FRUCT). 2017. pp. 148-155. DOI: 10.23919/FRUCT.2017.8250177

[12] Ghaffari E. Providing a new scheduling method in fog network using the ant colony algorithm. 2019 [Accessed: September 05, 2020]

[13] Bian S, Huang X, Shao Z. Online task scheduling for fog computing with multi-resource fairness. In: IEEE 90th Vehicular Technology Conference (VTC2019-Fall). 2019. pp. 1-5. DOI: 10.1109/VTCFall.2019.8891573

[14] Wang J, Li D. Task scheduling based on a hybrid heuristic algorithm for smart production line with fog computing. Sensors. 2019;**19**:5

[15] Li H, Ota K, Dong M. Deep reinforcement scheduling for mobile crowdsensing in fog computing. ACM Transactions on Internet Technology. 2019;**19**:21-28. DOI: 10.1145/3234463

[16] Wang S, Zhao T, Pang S. Task scheduling algorithm based on improved firework algorithm in fog computing. IEEE Access. 2020;**8**:32385-32394. DOI: 10.1109/ACCESS.2020.2973758

[17] Nikoui TS, Balador A, Rahmani AM, Bakhshi Z. Cost-aware task scheduling in

fog-cloud environment. In: 2020 CSI/
CPSSI International Symposium on
Real-time and Embedded Systems and
Technologies (RTEST). 2020. pp. 1-8.
DOI: 10.1109/RTEST49666.2020.9140118

[18] La QD, Ngo MV, Dinh TQ, Quek TQS,
Shin H. Enabling intelligence in fog
computing to achieve energy and latency
reduction. Digital Communication
Networking. 2019;5(1):3-9.
DOI: 10.1016/j.dcan.2018.10.008

[19] Heydari G, Rahbari D, Nickray M.
Energy Saving Scheduling in a Fog-
Based IoT Application by Bayesian Task
Classification Approach. 2019.
DOI: 10.3906/elk-1902-152

[20] Borthakur D, Dubey H, Constant N,
Mahler L, Mankodiya K. Smart fog: Fog
computing framework for unsupervised
clustering analytics in wearable Internet
of Things. In: 2017 IEEE Global
Conference on Signal and Information
Processing (GlobalSIP). 2017. pp.
472-476. Available from: http://bit.
ly/3TIw5LF

[21] Bittencourt LF, Diaz-Montes J,
Buyya R, Rana OF, Parashar M.
Mo-bility-aware application scheduling
in fog computing. IEEE Cloud
Computing. 2017;4(2):26-35.
DOI: 10.1109/MCC.2017.27

[22] Abdel-Basset M, Mohamed R,
Elhoseny M, Bashir AK, Jolfaei A,
Kumar N. Energy-aware marine
predators algorithm for task scheduling
in IoT-based fog computing
applications. IEEE Transaction on
Industrial Information. 2020:1-1.
DOI: 10.1109/TII.2020.3001067

[23] Rahbari D, Nickray M. Scheduling
of fog networks with optimized knapsack by symbiotic organisms search. In:
2017 21st Conference of Open
Innovations Association (FRUCT). 2017.
pp. 278-283. DOI: 10.23919/FRUCT.
2017.8250193

[24] Liu L, Qi D, Zhou N, Wu Y. A task
scheduling algorithm based on
classification mining in fog computing
environment. In: Wireless
Communications and Mobile Computing.
2018. DOI: 10.1155/2018/2102348

[25] Zhang Y, Chen X, Chen Y, Li Z,
Huang J. Cost efficient scheduling for
delay-sensitive tasks in edge computing
system. In: 2018 IEEE International
Conference on Services Computing
(SCC). 2018. pp. 73-80. DOI: 10.1109/
SCC.2018.00017

[26] Fang J, Chen Y, Lu S. A scheduling
strategy for reduced power consumption in mobile edge computing. In:
IEEE INFOCOM 2020-IEEE Conference
on Computer Communications
Workshops (INFOCOM WKSHPS).
2020. pp. 1190-1195. DOI: 10.3390/
app10176057

[27] Zhang Y, Xie M. A More Accurate
Delay Model based Task Scheduling in
Cellular Edge Computing Systems. 2019.
p. 76. DOI: 10.1109/ICCC47050.
2019.9064217

[28] Oo T, Ko Y-B. Application-aware
task scheduling in heterogeneous edge
cloud. In: 2019 International Conference
on Information and Communication
Technology Convergence (ICTC). 2019.
pp. 1316-1320. DOI: 10.1109/
ICTC46691.2019.8939927

[29] Fan J, Wei X, Wang T, Lan T,
Subramaniam S. Churn-resilient task
scheduling in a tiered IoT infrastructure.
China Communication. 2020:162-175

[30] Gamal H, Saroit IA, Kotb AM.
Grouped tasks scheduling algorithm
based on QoS in cloud computing
network. Egyptian Information Journal.
2017;18(1):11-19. DOI: 10.1016/j.
eij.2016.07.002

[31] Li G, Liu Y, Wu J, Lin D, Zhao S.
Methods of resource scheduling based

on optimized fuzzy clustering in fog computing. Sensors. 2019;**19**(9). DOI: 10.3390/s19092122

[32] Sami H, Mourad A, Otrok H, Bentahar J. FScaler: Automatic resource scaling of containers in fog clusters using reinforcement learning. In: 2020 International Wireless Communications and Mobile Computing (IWCMC). 2020. pp. 1824-1829. DOI: 10.1109/IWCMC48107.2020.9148401

[33] Tang C, Zhu C, Wei X, Chen W, Rodrigues JJPC. RSU-empowered resource pooling for task scheduling in vehicular fog computing. In: 2020 International Wireless Communications and Mobile Computing (IWCMC). 2020. pp. 1758-1763. DOI: 10.1109/IWCMC48107.2020.9148290

[34] Yang Y, Wang K, Zhang G, Chen X, Luo X, Zhou M-T. MEETS: Maximal energy efficient task scheduling in homogeneous fog networks. IEEE Internet Things Journal. 2018;**5**(5):4076-4087. DOI: 10.1109/JIOT.2018.2846644

[35] Liu Z, Yang X, Yang Y, Wang K, Mao G. DATS: Dispersive stable task scheduling in heterogeneous fog networks. IEEE Internet of Things Journal. 2019;**6**(2):3423-3436. DOI: 10.1109/JIOT.2018.2884720

[36] Yang M, Ma H, Wei S, Zeng Y, Chen Y, Hu Y. A multi-objective task scheduling method for fog computing in cyber-physical-social services. IEEE Access. 2020;**8**:65085-65095. DOI: 10.1109/ACCESS.2020.2983742

[37] Hoang D, Dang TD. FBRC: Optimization of task scheduling in fog-based region and cloud. In: 2017 IEEE Trustcom/BigDataSE/ICESS. 2017. pp. 1109-1114. DOI: 10.1109/Trustcom/BigDataSE/ICESS.2017.360

[38] Rahbari D, Nickray M. Task offloading in mobile fog computing by classification and regression tree.

Peer-Peer Network Application. 2020;**13**(1):104-122. DOI: 10.1007/s12083-019-00721-7

[39] Liu X et al. FogWorkflowSim: An Automated Simulation Toolkit for Workflow Performance Evaluation in Fog Computing. 2019. p. 1117. DOI: 10.1109/ASE.2019.00115

[40] Benblidia MA, Brik B, Merghem-Boulahia L, Esseghir M. Ranking fog nodes for tasks scheduling in fog-cloud environments: A fuzzy logic approach. In: 2019 15th International Wireless Communications Mobile Computing Conference (IWCMC). 2019. pp. 1451-1457. DOI: 10.1109/IWCMC.2019.8766437

[41] Javaid N, Butt AA, Latif K, Rehman A. Cloud and fog based integrated environment for load balancing using cuckoo levy distribution and flower pollination for smart homes. In: 2019 International Conference on Computer and Information Sciences (ICCIS). 2019. pp. 1-6. DOI: 10.1109/ICCISci.2019.8716467

[42] Apat HK. An optimal task scheduling towards minimized cost and response time in fog computing infra-structure. In: 2019 International Conference on Information Technology (ICIT). 2019. pp. 160-165

[43] Liu Z, Wang K, Li K, Zhou M-T, Yang Y. Parallel scheduling of multiple tasks in heterogeneous fog networks. In: 2019 25th Asia-Pacific Conference on Communications (APCC). 2019. pp. 413-418. DOI: 10.1109/APCC47188.2019.9026469

[44] Wang K, Zhou Y, Liu Z, Shao Z, Luo X, Yang Y. Online task scheduling and resource allocation for intelligent NOMA-based industrial internet of things. IEEE Journal on Selected Areas in Communications. May 2020;**38**(5): 803-815. DOI: 10.1109/JSAC.2020.2980908

[45] Yin L, Luo J, Luo H. Tasks scheduling and resource allocation in fog computing based on containers for smart manufacturing. IEEE Transactions on Industrial Information. 2018;**14**(10):4712-4721. DOI: 10.1109/TII.2018.2851241

[46] Mukherjee M, Guo M, Lloret J, Iqbal R, Zhang Q. Deadline-aware fair scheduling for offloaded tasks in fog computing with inter-fog dependency. IEEE Communications Letters. 2020;**24**(2):307-311. DOI: 10.1109/LCOMM.2019.2957741

[47] Xu J, Hao Z, Zhang R, Sun X. A method based on the combination of laxity and ant colony system for cloud-fog task scheduling. IEEE Access. 2019;7:116218-116226. DOI: 10.1109/ACCESS.2019.2936116

[48] Rahbari D, Kabirzadeh S, Nickray M. A security aware scheduling in fog computing by hyper heuristic algorithm. In: 2017 3rd Iranian Conference on Intelligent Systems and Signal Processing (ICSPIS). 2017. pp. 87-92. DOI: 10.1109/ICSPIS.2017.8311595

[49] Zhang G, Shen F, Zhang Y, Yang R, Yang Y, Jorswieck EA. Delay minimized task scheduling in fog-enabled IoT networks. In: 2018 10th International Conference on Wireless Communications and Signal Processing (WCSP). 2018. pp. 1-6. DOI: 10.1109/WCSP.2018.8555532

[50] Alzeyadi A, Farzaneh N. A novel energy-aware scheduling and load-balancing technique based on fog computing. In: 2019 9th International Conference on Computer and Knowledge Engineering (ICCKE). 2019. pp. 104-109. DOI: 10.1109/ICCKE48569.2019.8964946

[51] Bian S, Huang X, Shao Z, Yang Y. Neural task scheduling with reinforcement learning for fog computing systems. In: 2019 IEEE Global Communications Conference (GLOBECOM). 2019. pp. 1-6. DOI: 10.1109/GLOBECOM38437.2019.9014045

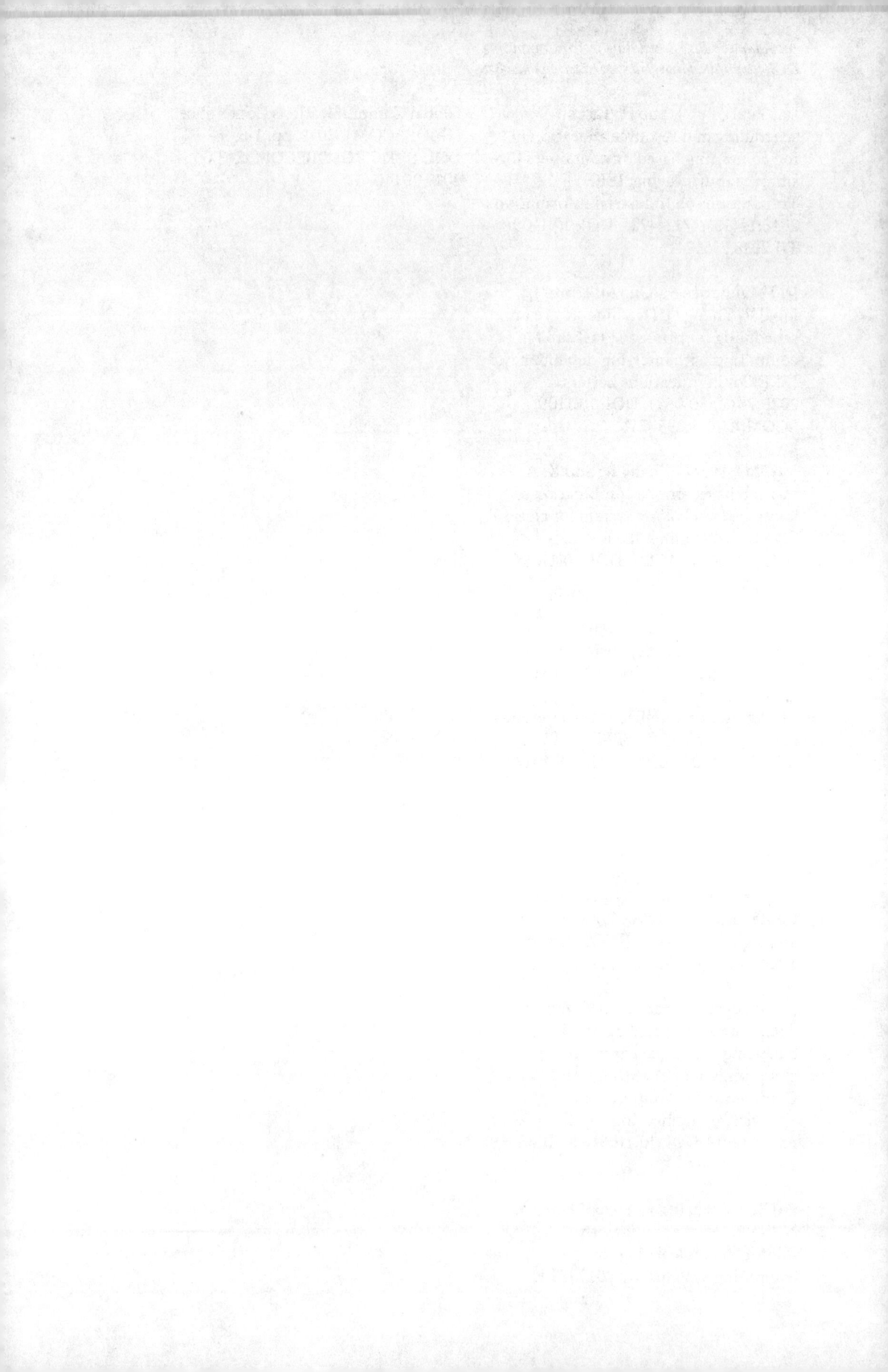

An Immune Multiobjective Optimization with Backtracking Search Algorithm Inspired Recombination

Hamed Ould Sidi, Rachid Ellaia, Emmanuel Pagnacco and Ahmed Tchvagha Zeine

Abstract

We propose a novel hybrid multiobjective (MO) immune algorithm for tackling continuous MO problems. Similarly to the nondominated neighbor immune algorithm (NNIA), it considers the characteristics of OM problems: based on the fitness values, the best individuals from the test population are selected and recombined to guide the rest of the individuals in the population to the Pareto front. But NNIA uses the simulated binary crossover (SBX), which uses the local search method. In our algorithm, the recombination is essentially inspired by the cross used in the backtracking search algorithm (BSA), but the adaptations are found in the immune algorithm. Thus, three variants are designed in this chapter, resulting in new recombination operators. They are evaluated through 10 benchmark tests. For the most advanced proposed variant, which is designed to have global search ability, results show that an improved convergence and a better diversity of the Pareto front are statistically achieved when compared with a basic immune algorithm having no recombination or to NNIA. Finally, the proposed new algorithm is demonstrated to be successful in approximating the Pareto front of the complex 10 bar truss structure MO problem.

Keywords: multiobjective optimization, evolutionary algorithms, backtracking search, hybrid recombination, artificial immune systems, truss optimization

1. Introduction

Planty of the multiobjective (MO) optimization problems lie in the engineering field. The objective functions are contradictory, the optimal solutions of these problems are known by the Pareto front. To get the optimal solutions of these problems, evolutionary algorithms (EAs) have been recognized to be very efficient in solving MO optimization problems by finding a representative Pareto front in one run. State-of-the-art algorithms are presented in [1–5]. These algorithms, which are population-based, are able to simultaneously explore various regions of the Pareto front.

In last past few years, immune systems have inspired new algorithms for resolutions OM problems. The fundamental principles of the artificial immune system (AIS) algorithm are clonal selection, [6] mutation, and more recently, recombination [7–15].

The nondominated neighbor-based immune optimization algorithm (NNIA) is effective to deal with MO problems [9]. NNIA has proved that it is advantageous to incorporate a crossover operator into the algorithm. To do this, it uses simulated binary crossover (SBX). But SBX is a recombination operator, which performs search near the recombination parent [16].

Backtracking Search Optimization Algorithm (BSA) is a new nature-inspired algorithm proposed by [17]. BSA's special mechanism to ensure a trial individual is effective, ability to learn fast solving different numerical optimization problems sequentially and quickly, with a clear structure. Since it was introduced, the BSA has attracted many research studies, and it has been applied to various optimization problems [18–20].

In this chapter, we develop a novel hybrid MO immune method to solve the problems of continuous multiobjective optimization. The NNIA algorithm uses the best individuals in the trial population to drive the search to Pareto front. But NNIA uses SBX, which mainly has local search capability. In our proposal, the recombination is inspired by the cross used in the BSA algorithm, but adaptations are found to fit the immune algorithm. Therefore, three variants are considered in the context of this chapter, which gives rise to new recombination operators for immune algorithm.

This chapter is organized as follows: In Section 2, we introduce the MO problem. In Sections 3 and 4, NNIA algorithm and BSA recombination are presented, and we propose new algorithms to solve the MO problem. The effectiveness of these algorithms is investigated in Section 5 when confronting to various MO test problems. In Section 6, the chosen algorithm is applied to solve the multiobjective topology optimization of truss structures. A short summary is proposed to conclude this paper.

2. Multiobjective optimization problem

The multiobjective optimization problem is formalized in this section. Concepts related to the Pareto front are introduced, firstly from a theoretical point of view and then when considering its approximation through a numerical approach [21].

2.1 Pareto front

Let us consider the following multiobjective optimization (MO) problem:

$$\min_{\mathbf{x} \in \Omega} \mathbf{f}(\mathbf{x}) = \left(f_1(\mathbf{x}), \cdots, f_m(\mathbf{x}) \right)^{\mathrm{T}} \tag{1}$$

where m is the number of objective functions, $\mathbf{x} = (x_1, \cdots, x_n) \in \Omega$ is the n_x-dimensional decision space where each decision variable x_i is bounded by lower and upper limits $x_{li} \leq x_i \leq x_{ui}$ for $i = 1, \cdots, n_x$.

Pareto-front-related concepts are [22]:

1. *Pareto dominance: Suppose $\mathbf{x}_a$ and $\mathbf{x}_b$ are two different feasible solutions to the MO problem. Then $\mathbf{x}_a$ dominates $\mathbf{x}_b$ if and only if*

$$f_i(\mathbf{x}_a) \leq f_i(\mathbf{x}_b)) \forall i \in \{1, \cdots, m\} \tag{2}$$

and:

$$\exists k \in \{1, \dots, m\} \ f_k(\mathbf{x}_a) < f_k(\mathbf{x}_b) \tag{3}$$

1. *Pareto-optimal solution:* A solution $\mathbf{x}^*$ is said to be Pareto-optimal if there does not exist another solution that dominates it.

2. *Pareto-optimal set:* The Pareto-optimal set is the set X^* of all Pareto-optimal solutions.

3. *Pareto-optimal front:* The Pareto-optimal front is the set F^* of values or outcomes of all the objective functions, which corresponds to the solutions:

$$F^* = \left\{ \mathbf{f}(\mathbf{x}^*) = \left(f_1(\mathbf{x}^*), \dots, f_m(\mathbf{x}^*) \right)^{\mathrm{T}} \ \text{such that}: \ \mathbf{x}^* \in X^* \right\} \tag{4}$$

2.2 Multiobjective solution

Otherwise, the following terms are cited in the reference [9]:

1. *Antibody*: An antibody refers to the coding of a decision variable $\mathbf{x}$. In this study, a real-valued representation is adopted, being $\mathbf{x}$ itself.

2. *Crowding distance*: The crowding distance (CD) is a measure for diversity maintenance [1]. It reads:

$$CD\left(\widehat{\mathbf{X}}\right) = \sum_{j=1}^{m} \frac{D_j\left(\widehat{\mathbf{X}}\right)}{f_j^{\max} - f_j^{\min} + \varepsilon_D} \tag{5}$$

where $f_j^{\max}$ and $f_j^{\min}$ are maximal and minimal values of the j-th objective respectively, ε_D is a small number and:

$$D_j\left(\widehat{\mathbf{X}}\right) = \begin{cases} \infty & \text{if } \ \widehat{\mathbf{x}}_k \text{ is a boundary point of } \widehat{\mathbf{X}} \\ \min \left| f_k\left(\widehat{\mathbf{X}}\right) - f_l\left(\widehat{\mathbf{X}}\right) \right| & \text{otherwise} \end{cases} \tag{6}$$

for $k, l \in \{1, \cdots, n_x\}$ such that $k \neq l \neq j$.

3. Immune optimization algorithm and recombination operator

3.1 Nondominated neighbor immune optimization algorithm

Nondominated neighbor immune algorithm (NNIA), using the nondominated neighbor-based selection and proportional cloning, pays more attention to the less-crowded regions of the current trade-off front.

We denote by $\widehat{\mathbf{X}}$ the dominant population, $\mathbf{X}_A$ the active population and $\mathbf{X}_C$ the clone population at time t are stored by time-dependent variable matrices, Their sizes are $n_{\widehat{x}}$, n_A, and n_C respectively.

The NNIA algorithm is presented in 1 where: = is the update operator. n_D, $n_{\widehat{X}\max}$, $n_{A\max}$, n_C, and n_{it}, the number of iterations.

Algorithm 1 Pseudo code of NNIA

Function $\widehat{X} = NNIA\left(n_x, m, \mathbf{f}(\mathbf{x}), \mathbf{x}_l, \mathbf{x}_u, n_{\widehat{X}\max}, n_{A\max}, n_C, n_{it}\right)$

1: Generate a uniform random initial population $\widehat{X}$ of size $n_C \times n_x$ in respect to $\mathbf{x}_l$ and $\mathbf{x}_u$;

2: $\widehat{X} := Find\_Non\_Dominated\left(\widehat{X}|\mathbf{f}(\mathbf{x})\right)$;

3: **for** $t = 0 : n_{it}$, **do**

4:    $X_A := CD\_Truncation\left(\widehat{X}, n_{A\max}\right)$;

5:    $X_C := Cloning(X_A, n_C)$;

6:    $X_C := Recombination(X_C, X_A, \mathbf{x}_l, \mathbf{x}_u)$;

7:    $X_C := Hypermutation(X_C, \mathbf{x}_l, \mathbf{x}_u)$;

8:    $\widehat{X} := Find\_Non\_Dominated\left(\left[\widehat{X}; X_C\right]|\mathbf{f}(\mathbf{x})\right)$;

9:    $\widehat{X} := CD\_Truncation\left(\widehat{X}, n_{\widehat{X}\max}\right)$;

end for

3.2 Recombination and crossovers

3.2.1 NNIA recombination

For a recombination, operation of NNIA has been adopted in many MO EAs [1, 4], an antibody of the cloning population and an antibody of the active population are selected and modified as:

$$\{X_C\}_{ij} := \begin{cases} \dfrac{1-\beta}{2}\{X_C\}_{ij} + \dfrac{1+\beta}{2}\{X_A\}_{kj} & \text{if} \quad a = 1 \;\&\; b = 0 \\[2mm] \dfrac{1+\beta}{2}\{X_C\}_{ij} + \dfrac{1-\beta}{2}\{X_A\}_{kj} & \text{if} \quad a = 1 \;\&\; b = 1 \\[2mm] \{X_C\}_{ij} & \text{if} \quad a = 0 \end{cases} \quad |a \sim \mathcal{U}(0,1),\; b \sim \mathcal{U}(0,1)$$

$$(7)$$

for $i \in \{1, \dots, n_C\}, j \in \{1, \dots, n_x\}$, and k a random integer uniformly chosen in $\{1, \dots, n_A\}$. Above, $\mathcal{U}$ is the uniform discrete distribution, and β is a sample from a random distribution having the density:

$$p(\beta) = \begin{cases} 0 & \text{if} \quad \beta < 0 \\[2mm] \dfrac{\eta+1}{2}\beta^{\eta} & \text{if} \quad 0 \leq \beta \leq 1 \\[2mm] \dfrac{\eta+1}{2\beta^{\eta+2}} & \text{if} \quad \beta > 1 \end{cases}$$

where η is a real nonnegative distribution. Hence, four independent random variables are involved in this recombination operation: a, b, k, and β. A boundary control is performed by:

$$\{\mathbf{X}_C\}_{ij} := \begin{cases} x_{li} & \text{if} \quad \{\mathbf{X}_C\}_{ij} < x_{li} \\ \{\mathbf{X}_C\}_{ij} & \text{if} \quad x_{li} \leq \{\mathbf{X}_C\}_{ij} \leq x_{ui} \\ x_{ui} & \text{if} \quad \{\mathbf{X}_C\}_{ij} > x_{ui} \end{cases}$$

3.2.2 Crossover operator of backtracking search optimization algorithm

Crossover strategy of BSA [17]. It consists in mixing two input populations $\mathbf{X}_P$ and $\mathbf{X}_Q$ to form a new output population $\mathbf{X}_R$, of equal sizes: $n_x \times n_x$. Then, BSA' crossover reads:

$$\{\mathbf{X}_R\}_{ij} := \begin{cases} \{\mathbf{X}_P\}_{ij} & \text{if} \quad \mathbf{T}_{ij} = 0 \\ \{\mathbf{X}_Q\}_{ij} & \text{if} \quad \mathbf{T}_{ij} = 1 \end{cases} \tag{8}$$

for $i \in \{1, ..., n_x\}, j \in \{1, ..., n_x\}$ and where $\mathbf{T}$ is a boolean matrix of sizes: $n_x \times n_x$, which is formed by following the algorithm 2.

To control the amount of mixing between the two populations $\mathbf{X}_P$ and $\mathbf{X}_Q$, we must define the parameter η such that $0 < \eta \leq n_x$. Moreover, we perform a random permutation on the lines of the $\mathbf{X}_P$ population before applying the relation (8).

Algorithm 2 Algorithm for the generation of the T matrix used in the BSA crossover

---

```
1: Initialize T := 0 and a := U(0, 1);
2: if a = 0 then
3:    for i = 1 : n_X do
4:       u := Permuting(1 : n_x);
5:       b := U(0, η);
6:       for k = 1 : b do
7:          j = u_k;
8:          T_ij = 1;
9:       end for
10:   end for
11: else
12:    for i = 1 : n_X do
13:       j := U(0, n_x);
14:       T_ij = 1;
15:    end for
16: end if
```

---

4. Recombination propositions for an hybrid algorithm

To get a more efficient immune algorithm, we will propose a hybridization method, which consists of exchanging the crossover operator used for recombination in NNIA with a new recombination operator inspired by BSA [23].

To find this new algorithm, we have to use two ideas:

1. The first idea consists of expanding the active population in order to obtain an extended active population, having its size equal to the clonal population size. The simplest way to achieve this consists of duplicating the active population;

2. The second idea consists of replacing the active population for the crossover by the clonal population, leading finally to a crossover that uses only the clonal population.

5. Experiments

In this section, we study the performance of the hybridization when solving some well-known MO techniques including five ZDT problems [24] and five DTLZ problems [25].

An optimization problem is typically written as:

$$\min_{\mathbf{x} \in \Omega} \quad \mathbf{f}(\mathbf{x}) = \left(f_1(\mathbf{x}), \cdots, f_m(\mathbf{x}) \right)^{\mathrm{T}} \tag{9}$$

where m is the number of objective functions, $\mathbf{x} = (x_1, \cdots, x_n) \in \Omega$ is the n_x-dimensional decision space where each decision variable x_i is bounded by lower and upper limits $x_{li} \leq x_i \leq x_{ui}$ for $i = 1, \cdots, n_x$.

5.1 Performance metrics

Approximate Pareto front solution of MO algorithms must achieve these two goals:

1. Convergence toward the true Pareto front; and

2. Diversity of solutions: the Pareto front must be uniformly distributed and spread over the entire feasible objective space to adequately capture the trade-offs.

For benchmark test problems, the true Pareto front is known, allowing to exploit performance metrics that used it.

We opted for two performance metrics for assessing algorithms efficiency. To measure the extent of the convergence to the true set of Pareto-optimal solutions and the spread of the Pareto front set, a normalized version of the inverted generational distance (IDG) metric proposed by [26] is adopted, while a normalized version of the spacing metric introduced by [27] enables to measure the uniformity of the obtained solutions.

5.1.1 Normalized inverted generational distance

The normalized inverted generational distance (NIGD) is based on a proposition of [26]. For measuring of the distance between the true Pareto front F^*, which is known at n^* discrete values—and stored in the matrix $\mathbf{F}(\mathbf{X}^*)$—and approximate solutions of the Pareto-optimal front $\mathbf{F}\left(\widehat{\mathbf{X}}\right)$:

$$\mathrm{NIGD}\left(\mathbf{F}(\mathbf{X}^*), \mathbf{F}\left(\widehat{\mathbf{X}}\right)\right) = \frac{1}{n^*} \sqrt{\sum_{j=1}^{n^*} c_j^2}, \tag{10}$$

for:

$$c_j = \min_{i \in \{1, \, \dots, \, n_x\}} \left(\sqrt{\sum_{k=1}^{m} \left(\overline{F}_k(\mathbf{X}_i^*) - \overline{F}_k\left(\widehat{\mathbf{X}}_j\right) \right)^2} \right), \qquad j \in \{1, \dots, n^*\}$$

where $\bar{\bullet}$ denotes a normalized objective function, ranging from 0 to 1 and defined by:

$$\overline{F}_k(\mathbf{x}) = \frac{F_k(\mathbf{x}) - \min\left(F_k(\mathbf{X}^*)\right)}{\max\left(F_k(\mathbf{X}^*)\right) - \min\left(F_k(\mathbf{X}^*)\right)}. \tag{11}$$

To obtain smaller values of this measure, the approximated set $F\left(\hat{\mathbf{X}}\right)$ must be very close to the Pareto front and cannot miss any part of the whole Pareto front at the same time.

5.1.2 Normalized spacing measure

The spacing metric introduced by [27] is modified by taking normalized objectives functions. This leads to the normalized NSP measure, defined by:

$$\text{NSP}(\mathbf{F}(\hat{\mathbf{X}})) = \sqrt{\frac{1}{n_X - 1}\sum_{j=1}^{n_X}(\mathbf{d}_j - \overline{d})^2} \tag{12}$$

for:

$$\mathbf{d}_j = \min_{\substack{i \in \{1, \ldots, n_X\} \\ i \neq j}} \left(\sqrt{\sum_{k=1}^{m}\left(\overline{F}_k(\hat{\mathbf{X}}_i) - \overline{F}_k(\hat{\mathbf{X}}_j)\right)^2}\right), \quad j \in \{1, \ldots, n_X\}$$

where $\overline{d}$ is the mean of $\mathbf{d}$.

5.2 Empirical comparison

In this section, performance of five NNIA variants are evaluated. The five variants are:

1. NNIA-X: the NNIA algorithm without crossover;

2. NNIA: the algorithm proposed by [9];

3. NNIA+X1: the hybridization of the NNIA algorithm with the BSA crossover by using the first strategy proposed in the Section 4. Inputs of the BSA crossover function are (1) the clonal population and (2) a random permutation of an extended active population obtained by duplicating individuals;

4. NNIA+X2: the hybridization of the NNIA algorithm with the BSA crossover by using the second strategy proposed in the Section 4. Inputs of the BSA crossover function are (1) the clonal population and (2) a random permutation of the clonal population;

5. NNIA+X3: the hybridization of the NNIA algorithm with the BSA crossover by using the third strategy proposed in the Section 4. Inputs of the BSA crossover function are (1) the clonal population and (2) a random permutation of an extended active population obtained by duplicating individuals with a proportion of random individuals.

For NNIA, parameters proposed in Ref. [9] are set:

- maximum size of dominant population: $n_{\hat{X} \max} = 100$,

- maximum size of active population: $n_{A \max} = 20$, and

- size of clone population $n_C = 100$,

with the distribution index for SBX that is 15, the distribution index for polynomial mutation that is 20 and the mutation probability of $1/n_x$ and the number of iterations stopped at 250. For NNIA+X3, the proportion of random individuals is chosen to be equal to n_A and their distribution is uniform.

Figures 1 and **2** show the statistic box plots for NIGD and NSP obtained for 1000 independent runs performed on each test problems ZDT and DTLZ that are chosen by [9][1].

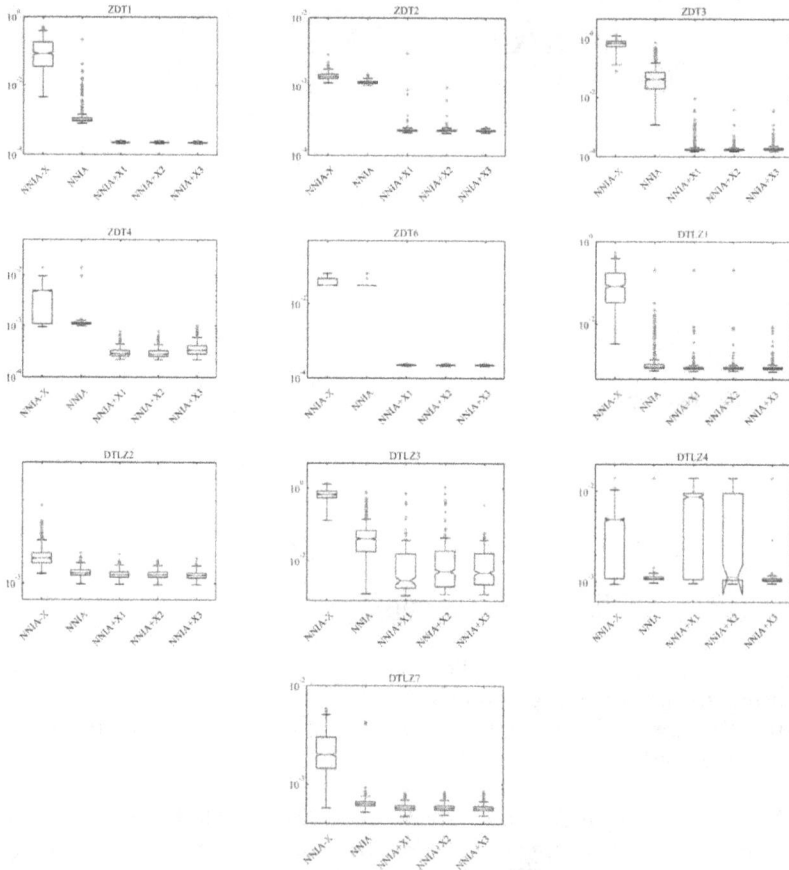

Figure 1.
NIGD obtained from 1000 independent runs of problems ZDT1, ZDT2, ZDT3, ZDT4, ZDT6, DTLZ1, DTLZ2, DTLZ3, DTLZ4, and DTL7.

---

[1] From informations given in [25], it is believed that the problem denoted DTLZ6 in [9] is in fact the DTLZ7 problem.

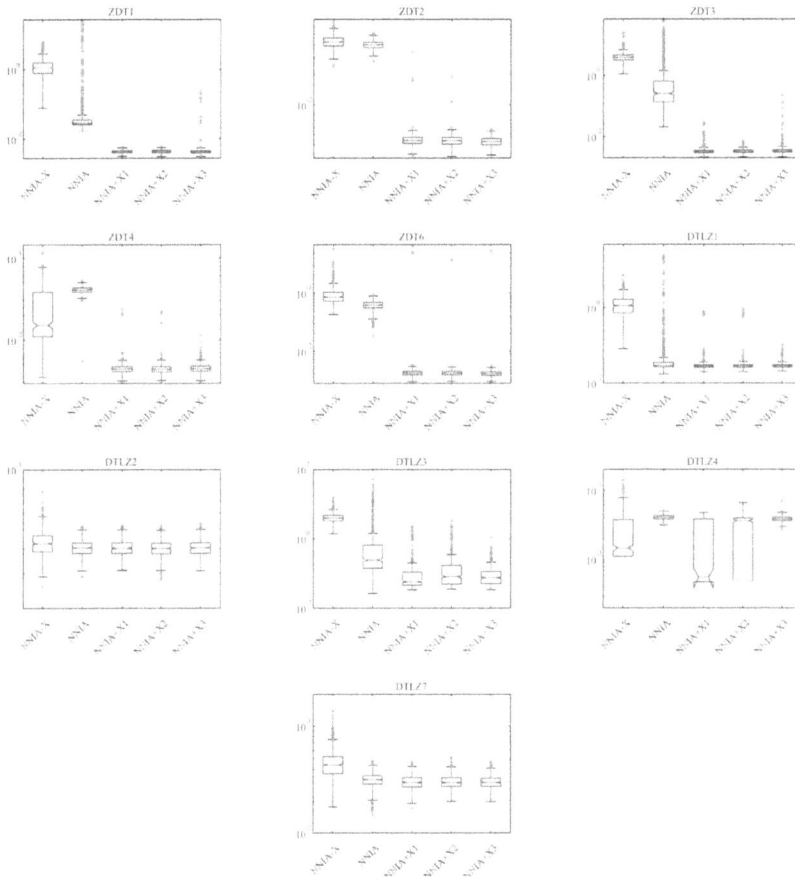

Figure 2.
Statistics box plots of NSP obtained from 1000 independent runs of benchmark test problems ZDT1, ZDT2, ZDT3, ZDT4, ZDT6, DTLZ1, DTLZ2, DTLZ3, DTLZ4, and DTL7.

NIGD's statistical results show that the NNIA algorithm is better than NNIA-X for the problems addressed. We also observe the efficiency of the NNIA+X1 and NNIA +X2 algorithms compared with NNIA with the exception of the difficult DTLZ4 test problem. For ZDT4 problem, NNIA+X3 is lower than NNIA+X1 and NNIA+X2. But we can always notice that our proposed algorithm NNIA+X3 remains superior to NNIA for the problems treated. Except for the two issues ZDT4 and DTLZ4, NSP shows the superiority of NNIA over NNIA-X. In all treated cases, NNIA+X1, NNIA +X2, and NNIA+X3 appear to be equal to or greater than NNIA. But for all these algorithms, the DTLZ4 problem seems to be the most difficult, since there are runs for which the Pareto front is approximated by a single, unique, point.

Table 1 shows the percentage of results showing a single point for the Pareto front of the DTLZ4 test problem when a sequence of 1000 runs is performed with

NNIA-X	NNIA	NNIA+X1	NNIA+X2	NNIA+X3
2%	11%	13%	11%	0.2%

Table 1.
Percentage of results exhibiting a single point for the Pareto front of the DTLZ4 test problem when 1000 runs are carrying out.

each algorithm. Generally, we conclude that NNIA+X3 retains better population diversity, and its convergence is faster than NNIA for these two and three objective test problems.

6. Experiments on the 10 bar truss design problem

In this section, we address the multi-objective sizing optimization of truss-like structures which is a continuous subject of researches in structural design [28–30].

6.1 Problem formulation

In this study, we consider the 10 bar truss test, ketch in the **Figure 3**. Two objective functions have to be minimized: the mass and the displacement; and one objective function has to be maximized: the first flexible natural frequency of the structure.

Denoting $\mathbf{x} \in \Omega$ the vector of the topological and sizing optimization parameters, such that $0 \leq x_i \leq 1$ for $i \in \{1, \dots, n\}$ where $n = 10$ is the number of elements, the three individual objectives are:

    1. The mass w of the structure

$$w(\mathbf{x}) = \sum_{i=1}^{n} \rho A l_i x_i,$$

where l_i is the length of the i-th element, $\rho = 2768$ kg/m^3 is the density of the material and $A = 0.01419352$ m^2 is the element cross-section area.

    2. The maximum displacement u of the structure

$$u(\mathbf{x}) = \max \left(\mathbf{u}^* = \arg \min_{\mathcal{S}} \left(\frac{1}{2} \mathbf{u}^T \mathbf{K}(\mathbf{x}) \mathbf{u} - \mathbf{u}^T \mathbf{F} \right) \right),$$

where:

- **F** is the vector of loads

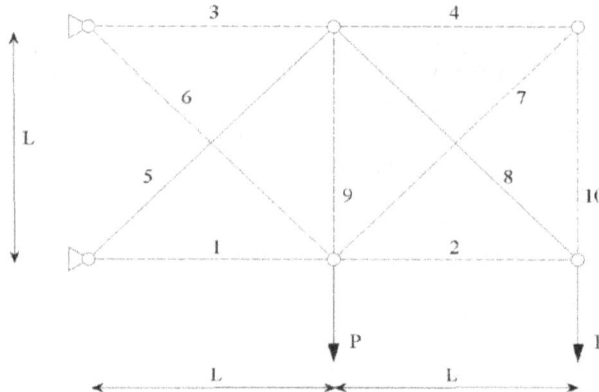

Figure 3.
Sketch of the 10 bar truss.

- **K** is the stiffness matrix of the finite element (FE) model, having the Young' modulus $E = 68.95$ GPa.

The set S refers to the kinematic admissible space, *i.e.* the one that satisfies the imposed boundary conditions given by the supports while carrying all the prescribed loads, where $P = 448.2$ kN.

3. The function (minimum flexible natural frequency f) to maximize it

$$-f(\mathbf{x}) = -\min\left(\frac{1}{2\pi}\omega^*\right),$$

$$\text{where}: \quad \{\omega^{*2}, \mathbf{u}^*\} = \arg\min_{\mathbf{u} \in S}\left(\omega^2 = \frac{\mathbf{u}^T \mathbf{K}(\mathbf{x})\mathbf{u}}{\mathbf{u}^T \mathbf{M}(\mathbf{x})\mathbf{u}}\right), \quad \|\mathbf{u}\| \neq 0$$

where **M** is the mass matrix of the FE model[2].

Moreover, this MO problem is subjected to constraints for the mechanical stress σ_i for each element i:

$$|\sigma_i(\mathbf{x})| \leq \bar{\sigma} \qquad i \in \{1, \dots, n\}$$

where $\bar{\sigma} = 172.4$ MPa is the yield strength.

As designs with local rigid body modes or kinematic modes are not of interest, constraints are added to the MO problem formulation:

$$\frac{|\sigma_i(\mathbf{x})|}{\bar{\sigma}} > \varepsilon, \qquad i \in \{1, \dots, n\} \text{ such that } x_i > 0$$

where $\varepsilon = 0.001$.

Since the optimal Pareto front is unknown for this problem, unnormalized metric indicators are to assess for the MO algorithm performance. Thus, in practice, we introduce an *a priori* scaling of the three objectives, by defining:

$$f_1(\mathbf{x}) = \frac{w(\mathbf{x})}{7,000}, \qquad f_2(\mathbf{x}) = \frac{u(\mathbf{x}) - 0.016}{20}, \qquad f_3(\mathbf{x}) = \frac{22,500 - (2\pi f(\mathbf{x}))^2}{5,000}$$

Moreover, in order to handle constraints of this MO problem, we use the penalty method. This technique consists of replacing the constrained optimization problems by an optimization problems without constraints, when introducing new objective functions to be optimized:

$$\phi_k(\mathbf{x}) = f_k(\mathbf{x}) + r\varphi(\mathbf{x}) \tag{13}$$

where the penalty function chosen here is:

$$\varphi(\mathbf{x}) = \sum_{i=1}^{n}\left(\max\left\{0, \frac{|\sigma_i(\mathbf{x})|}{\bar{\sigma}} - 1\right\}\right)^2 + \sum_{i=1}^{n}\left(\max\left\{0, \varepsilon - \frac{|\sigma_i(\mathbf{x})|}{\bar{\sigma}}\right\}\right)^2 \tag{14}$$

and where r is a positive penalty parameter. We have chosen here $r = 10^{10}$.

---

[2] To obtain the best numerical efficiency for the FE analysis, the FE disassembly strategy proposed in Ref. [31] is involved.

Finally, the MO problem definition for the 10 bar truss of this work is:

$$\min_{\mathbf{x} \in \Omega} \left(f_1(\mathbf{x}) + r\varphi(\mathbf{x}),\, f_2(\mathbf{x}) + r\varphi(\mathbf{x}),\, f_3(\mathbf{x}) + r\varphi(\mathbf{x}) \right)$$

6.2 Numerical simulations for two objective functions

In this subsection, we will subdivide and transform the 10 bar MO problem from the previous section into three MO subproblems. Objective functions are considered two by two: (w,u), (w,f), and (u,f) To solve each of these 10 bar MO problems, we use the NNIA algorithms and the NNIA+X3, keeping the parameters to those of the previous subsection 5.2.

After 250 and 750 iterations, we obtain the two **Figures 4** and **5** (respectively), which show Pareto fronts of a typical execution, if the two algorithms start from the same initial population. In these figures, we observe that the NNIA+X3 algorithm shows better diversity for each subproblem, and that NNIA+X3 gives better convergence for the subproblems (w,f) and (u,f). Since each iteration of one of these algorithms requires $n_c = 100$ evaluations of the mechanical problem, 25,000 function evaluations are performed when 250 iterations are performed, and 75,000 function evaluations are performed when 750 iterations are performed.

Figure 6 shows the evolution of two metric indicators along the number of iterations for one typical run. Metric indicators chosen here are spacing and hyper-volume of Pareto fronts. Spacing evolution is presented in log-log scale in the figure. Each evaluation of the hyper-volume is achieved by using the same anti-utopia point and utopia point for results consistency. Moreover, in order to compare the three MO results on the same graph, a relative hyper-volume is plotted: the graph corresponds to the hyper-volume obtained divided by its maximum value. These graphs show a better diversity and convergence for NNIA+X3 compared with NNIA when early number of iterations are considered.

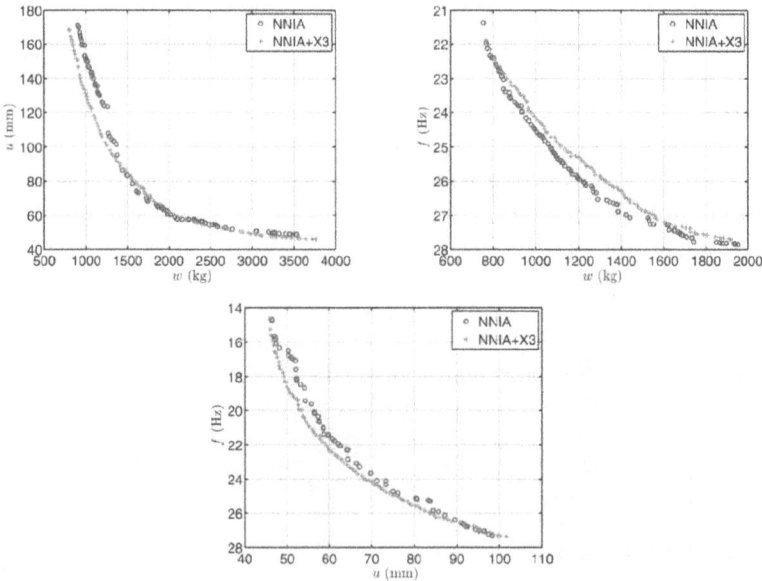

Figure 4.
Pareto fronts of the 10 bar truss MO problem two by two: (w,u) (up-left), (w,f) (up-right), (u,f) (down), after 250 iterations.

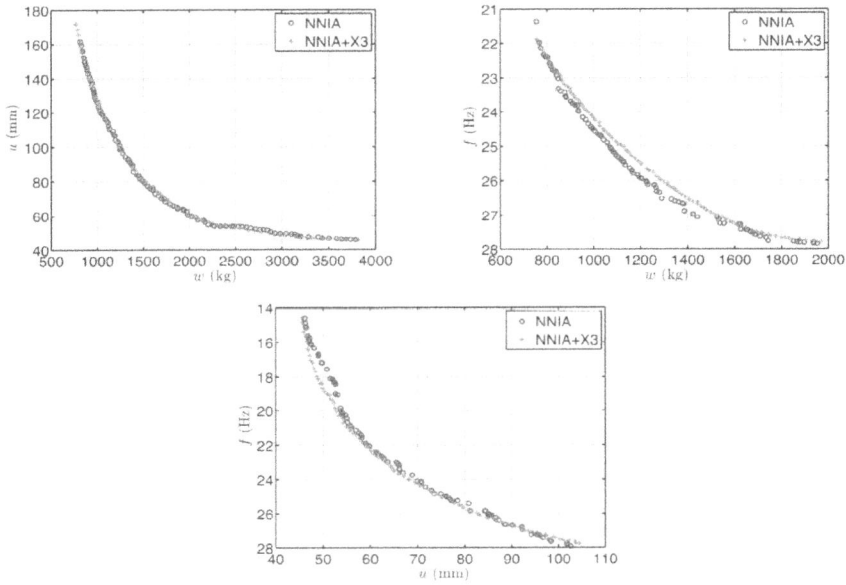

Figure 5.
Pareto fronts of the 10 bar truss MO problem, after 750 iterations for two by two: (w,u) (up-left), (w,f) (up-right), (u,f) (down).

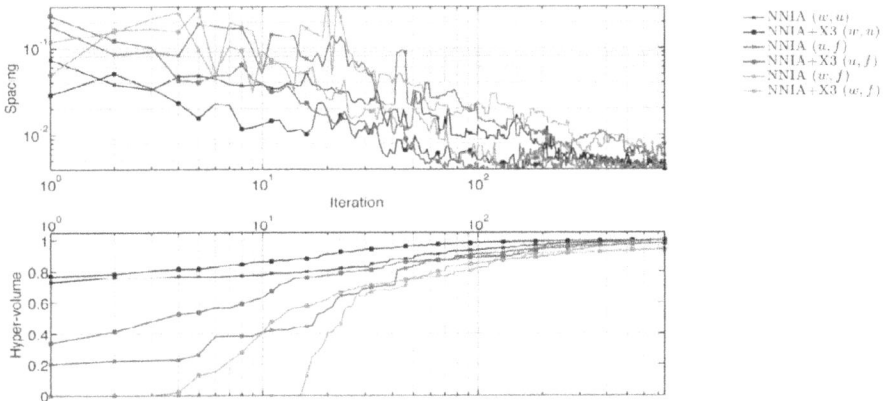

Figure 6.
Metrics indicators of the 10 bar truss MO problem for the three objectives functions two by two: spacing (up) and relative hyper-volume (down).

Tendency observed in the previous figure is confirmed by statistical results of **Figure 7** and **Figure 8**. These figures show box plots statistic for spacing and hyper-volume (respectively) when 300 runs stopped at 250 iterations are carried out. For spacing, means and variance are clearly better for NNIA+X3. Hyper-volume statistic results are also better for the NNIA+X3 when considering the (w,u) and (u,f) subproblems, while they are almost identicals for the (w,f) supbproblem, although the mean and variance are also better for the NNIA+X3. From results for the hyper-volume of the (w,f) supbproblem, it is assessed that this subproblem is the most difficult to solve since a wide spread is observed in data for both algorithms.

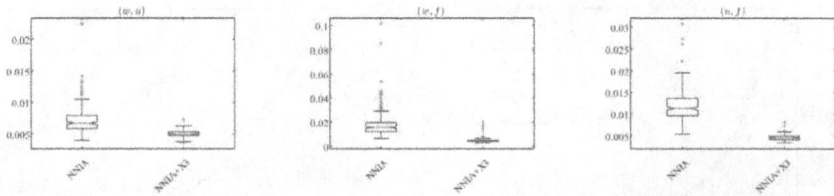

Figure 7.
Statistics box plots of spacing for 300 runs of the two-by-two MO 10 bar subproblems: (w, u) (left), (w, f) (middle), (u, f) (right).

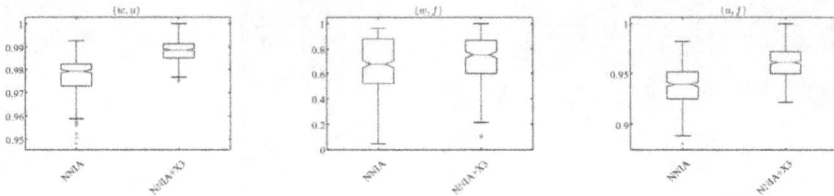

Figure 8.
Statistics box plots of relative hyper-volume for 300 runs of the two-by-two MO 10 bar subproblems: (w, u) (left), (w, f) (middle), (u, f) (right).

6.3 Numerical simulation for three objective functions

Figure 9 shows different views of the Pareto front obtained for one typical run when solving the three objectives 10 bar truss problem, using NNIA+X3 with the following parameter values: size of active population 30, clonal scale 150 and 750 iterations. In this case, the size of the dominant population is not limited to any number and all Pareto points found are kept. **Figure 10** shows the evolution of the

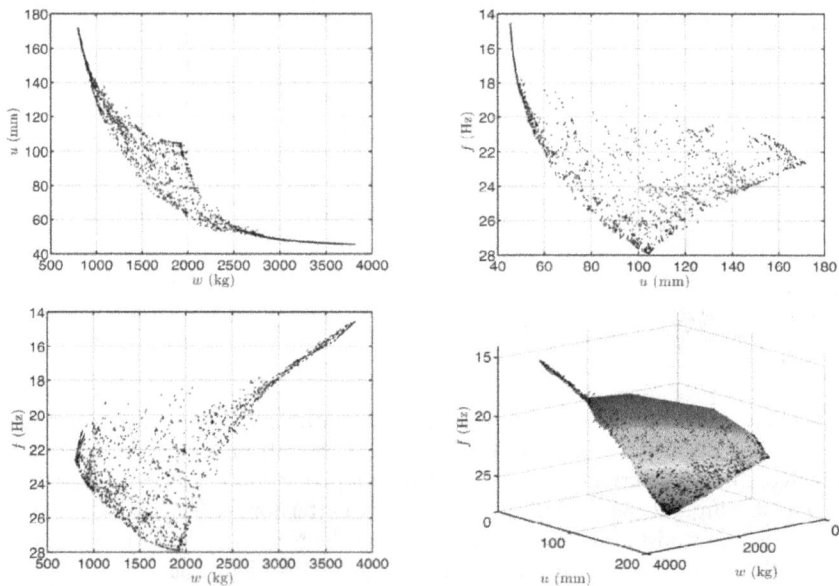

Figure 9.
Four different views of the Pareto front obtained for solving the three objective functions of the 10 bar truss problem; Colorized surface of the down-right subfigure is added for a better visualization, and the color corresponds to the frequency objective f.

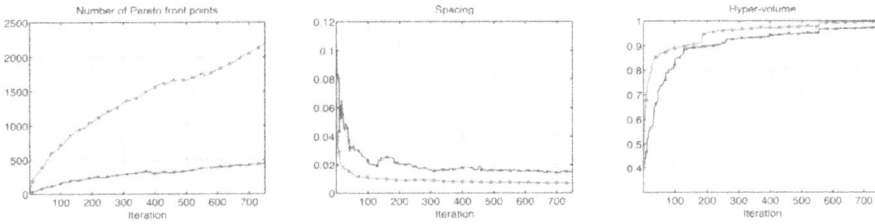

Figure 10.
Evolution of results for a typical run of the MO 10 bar problem: Number of points for the Pareto front (left), spacing (middle), and relative hyper-volume (right); Blue line with cross markers: NNIA; Red line with squared markers: NNIA+X3.

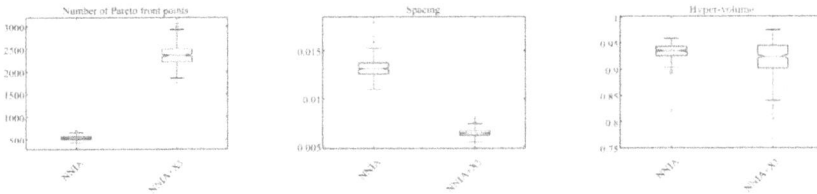

Figure 11.
Statistics box plots for 300 runs of the three objectives 10 bar problem with NNIA and NNIA+X3 with random initial population: Number of Pareto front points (left), spacing (middle), and relative hyper-volume (right).

number of points in the dominant population for the Pareto front given in **Figure 9**. It ends at 2216 Pareto points for this run.

Figure 11 shows box plot statistics when 300 runs are carried out with NNIA and NNIA+X3. It is observed that the number of points for the Pareto front is higher for NNIA+X3, with a better spacing. But the hyper-volume is better for NNIA. Detailled analysis of results has revealed that bad results for hyper-volume are due to a slow convergence to an extreme Pareto front point: the individual optima for the frequency objective. For this problem, the individual minima found for the frequency objective are most of the time better for NNIA than for NNIA+X3. However, it is also found that individual minima of the three objectives are rarely found in the Pareto front by both algorithms.

For better results, the idea is to handle the three individual minima found by a mono-objective optimization into the random initial population of both NNIA and NNIA+X3. This simple modification greatly improves performance results. **Figure 12** shows statistics box plots when 300 runs of MO problem are carried out when the three individual optima are given in the initial population. In such a situation and for each of the 300 runs done, NNIA+X3 appears to be superior to NNIA for all performance aspects, including the computed hyper-volume.

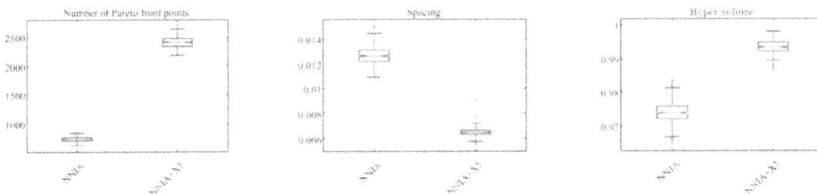

Figure 12.
Statistics box plots for 300 runs of the three objectives 10 bar problem with NNIA and NNIA+X3 when individual optima are handled in the initial population: Number of Pareto front points (left), spacing (middle), and relative hyper-volume (right).

7. Summary

This work is devoted to recombination for an NNIA algorithm. We propose three recombinations, inspired by the BSA algorithm crossing operator when adapting input populations.

In the first NNIA+X1 algorithm, the clonal population and an extended active population are concerned, when the extended active population is founded by duplicating individual antibodies.

In the second algorithm, NNIA+X2, recombination is achieved by using the clonal population and itself.

The NNIA+X3 algorithm uses the clonal population and an extended working population, which finds by duplicating individual antibodies and a proportion of random individuals. From this algorithm, a certain degree of mutation is carried out. The results obtained for the benchmark, ZDT, and DTLZ functions show that our proposed algorithm NNIA+X3 can accelerate the speed of convergence and maintain the desirable diversity, especially when solving problems with many local Pareto-optimal fronts. The experimental results of this algorithm to solve the problems of bi-objectives and three-objectives of optimization of 10 bar trellis structure indicate that the proposed NNIA+X3 surpasses the NNIA algorithm in terms of convergence rate and of course of quality of the solution.

Author details

Hamed Ould Sidi[1], Rachid Ellaia[2], Emmanuel Pagnacco[3]
and Ahmed Tchvagha Zeine[1*]

1 URMCD, University of Nouakchott Al Aasriya, Nouakchott, Mauritania

2 LERMA-EMI, Mohammadia School of Engineers, Rabat, Morocco

3 LMN, INSA de Rouen Normandie, St. Etienne du Rouvray, France

*Address all correspondence to: nouneahmed@gmail.com

IntechOpen

References

[1] Deb K, Pratap A, Agarwal S, Meyarivan T. A fast and elitist multiobjective genetic algorithm: NSGA-II. IEEE Transactions on Evolutionary Computation. 2002;6(2): 182-197

[2] Sierra MR, Coello CAC. Multi-objective particle swarm optimizers: A survey of the state-of-the-art. International Journal of Computational Intelligence Research. 2006;2:287-308

[3] Song MP, Gu GC. Research on particle swarm optimization: A review. In: Proceedings of the International Conference on Machine Learning and Cybernetics. Vol. 4. 2004. pp. 2236-2241

[4] Zitzler E, Laumanns M, Thiele L. SPEA2: Improving the strength Pareto evolutionary algorithm. In: Evolutionary Methods for Design, Optimization and Control with Applications to Industrial Problems. Athens, Greece; 2002. pp. 95-100

[5] Zitzler E, Thiele L. Multiobjective evolutionary algorithms: A comparative case study and the strength Pareto approach. IEEE Transactions on Evolutionary Computation. 1999;3(4): 257-271

[6] Omkar SN, Khandelwal R, Yathindra S, Naik GN, Gopalakrishn S. Artificial immune system for multi-objective design optimization of composite structures. Engineering Applications of Artificial Intelligence. 2008;21: 1416-1429

[7] Coello C, Cortes N. Solving multiobjective optimization problems using an artificial immune system. Genetic Programming and Evolvable Machines. 2005;6(2):163-190

[8] Gao J, Wang J. WBMOAIS: A novel artificial immune system for multiobjective optimization. Computers

and Operations Research. 2010;37(1): 50-61

[9] Gong M, Jiao L, Du H, Bo L. Multiobjective immune algorithm with nondominated neighbor-based selection. Evolutionary Computation. 2008;16(2):225-255

[10] Jiao L, Liu F, Ma W. A novel immune clonal algorithm for MO problems. IEEE Transactions on Evolutionary Computation. 2012;16(1): 35-50

[11] Luh GC, Chueh CH, Liu WW. MOIA: Multi-objective immune algorithm. Engineering Optimization. 2003;35:143-164

[12] Shang R, Jiao L, Liu F, Ma M. A novel immune clonal algorithm for MO problems. IEEE Transactions on Evolutionary Computation. 2012;16(1): 35-50

[13] Shi J, Gong M, Ma W, Jiao L. A multiobjective immune algorithm based on a multiple-affinity model. European Journal of Operational Research. 2010; 202(1):60-72

[14] Zinflou A, Gagn C, Gravelc M. GISMOO: A new hybrid genetic/immune strategy for multiple-objective optimization. Computers and Operations Research. 2012;9(9): 1951-1968

[15] Qi YT, Hou ZT, Yin ML, Sun HL, Huang JB. An immune multi-objective optimization algorithm with differential evolution inspired recombination. Applied Soft Computing. 2015;547(29): 395-410

[16] Deb K, Beyer HG. Self-adaptive genetic algorithms with simulated binary crossover. Evolutionary Computation. 2001;9(2):197-221

[17] Civicioglu P. Backtracking search optimization algorithm for numerical optimization problems. Applied Mathematics and Computation. 2013; **219**(15):8121-8144

[18] Sheoran Y, Kumar V, Rana KPS, Mishra P, Kumar J, Nair SS. Development of backtracking search optimization algorithm toolkit in LabVIEW. Procedia Computer Science. 2015;**57**:241-248

[19] Civicioglu P. Circular antenna array design by using evolutionary search algorithms. Progress In Electromagnetics Research B. 2013;**54**: 265-284

[20] Chaib AE, Bouchekara HREH, Mehasni R, Abido MA. Optimal power flow with emission and non-smooth cost functions using backtracking search optimization algorithm. International Journal of Electrical Power and Energy Systems. 2016;**81**:64-77

[21] Fang S, Yunfang C, Weimin W. Multi-objective optimization immune algorithm using clustering. Computing and Intelligent Systems. 2011;**234**: 242-251

[22] Bosman PAN, Thierens D. The balance between proximity and diversity in multiobjective evolutionary algorithms. IEEE Transactions on Evolutionary Computation. 2003;7(2): 174-188

[23] Chen J, Lin Q, Ji Z. A hybrid immune multiobjective optimization algorithm. European Journal of Operational Research. 2010;**204**(2): 294-302

[24] Zitzler E, Deb K, Thiele L. Comparison of multiobjective evolutionary algorithms: Empirical results. Evolutionary Computation. 2000;**8**(2):173-195

[25] Deb K, Thiele L, Laumanns M, Zitzler E. Scalable Multi-Objective Optimization Test Problems. Technical Report 112. Zurich, Switzerland: Computer Engineering and Networks Laboratory (TIK), Swiss Federal Institute of Technology (ETH); 2001. epagnacc, 2016.07.21

[26] Sierra MR, Coello CAC. Improving PSO-based multi-objective optimization using crowding, mutation and ε-dominance. In: Proceedings of the Evolutionary Multi-Criterion Optimization. Vol. 3239. 2005. pp. 263-276

[27] Schott JR. Fault tolerant design using single and multicriteria genetic algorithm optimization [masters thesis]. Dept. Aeronautics and Astronautics, Massachussets Institue of Technology; 1995

[28] Arkadiusz M. Geometrical aspects of optimum truss like structures for three-force problem. Structural and Multidisciplinary Optimization. 2012; **45**(1):21-32

[29] Richardson JN, Adriaenssens S, Bouillard P, Coelho RF. Multiobjective topology optimization of truss structures with kinematic stability repair. Structural and Multidisciplinary Optimization. 2012;**46**:513-532

[30] Kaveh A, Laknejadi K. A hybrid evolutionary graph-based multi-objective algorithm for layout optimization of truss structures. Acta Mechanica. 2013;**224**(2):343-364

[31] Hemez FM, Pagnacco E. Statics and inverse dynamics solvers based on strain-mode disassembly. Revue Européenne des Eléments Finis. 2000; **9**(5):511-560

Metaheuristics for Traffic Control and Optimization: Current Challenges and Prospects

Arshad Jamal, Hassan M. Al-Ahmadi,

Farhan Muhammad Butt, Mudassir Iqbal,

Meshal Almoshaogeh and Sajid Ali

Abstract

Intelligent traffic control at signalized intersections in urban areas is vital for mitigating congestion and ensuring sustainable traffic operations. Poor traffic management at road intersections may lead to numerous issues such as increased fuel consumption, high emissions, low travel speeds, excessive delays, and vehicular stops. The methods employed for traffic signal control play a crucial role in evaluating the quality of traffic operations. Existing literature is abundant, with studies focusing on applying regression and probability-based methods for traffic light control. However, these methods have several shortcomings and can not be relied on for heterogeneous traffic conditions in complex urban networks. With rapid advances in communication and information technologies in recent years, various metaheuristics-based techniques have emerged on the horizon of signal control optimization for real-time intelligent traffic management. This study critically reviews the latest advancements in swarm intelligence and evolutionary techniques applied to traffic control and optimization in urban networks. The surveyed literature is classified according to the nature of the metaheuristic used, considered optimization objectives, and signal control parameters. The pros and cons of each method are also highlighted. The study provides current challenges, prospects, and outlook for future research based on gaps identified through a comprehensive literature review.

Keywords: metaheuristics, intelligent traffic control, signal optimization, swarm intelligence, evolutionary computation, transport networks

1. Introduction

1.1 Traffic congestion: a challenging front

Recent decades have witnessed a rapid surge in population growth. Consequently, a high concentration of social and economic activities in urban metropolitans has led to the emergence of various transportation modes and services. Urban traffic congestion has become a daunting challenge in cities around the world. Excessive delay, low traveling speeds, increased travel costs, elevated drivers' anxiety and frustrations, high fuel consumption, and vehicular emissions are the few consequences of traffic congestion. It also poses a threat to a stable urban economy [1, 2]. Traffic demands fluctuate significantly during the day (TOD), especially during rush hours, which is one of the main causes of congestion buildup. Congestion may be recurrent, arising from routine cyclic fluctuations in traffic volumes, or it may be non-recurrent produced due to unforeseen events such as traffic incidents, unpredictable weather conditions. Existing transport infrastructure cannot withstand the ever-growing traffic demands, while the inappropriate allocation of temporal and spatial resources further exacerbates the problems [3, 4]. An effective solution to mitigate traffic congestion is to embed intelligent transportation system (ITS) technologies in existing transport infrastructure for efficient and sustainable operations. Researchers and practitioners have proposed various strategies such as signal control optimization and dynamic lane grouping to address the issue in recent years.

1.2 Traffic signal control (TSC)

Signalized intersections are a vital component of urban traffic networks and play a pivotal role in traffic control and management strategies. Over the years, they have been the primary focus of traffic improvement efforts since they are representative of frequent and restrictive bottlenecks. Poor traffic management at urban intersections leads to traffic jams and unsustainable travel patterns network-wide. Alternatively, intelligent traffic control and better management at these critical locations could result in smooth, safe, cheap, and sustainable operations. Traffic Signal Control (TSC) is an integral part of ITS. TSC is an important operation that can tackle various urban traffic issues such as congestion, fuel consumption and exhaust emission, and inefficient resource utilization. TSC involves determining appropriate signal timings parameters to improve various traffic performance measures like average vehicle delay, travel time, maximizing throughput, and reducing queue lengths and vehicular emissions. One of the main objectives of traffic signal control is to facilitate the safe and efficient movement of people through a road network. Achieving this goal warrant establishment of an accommodation plan that ensures appropriate assignment of right-of-way (ROW) to different users.

1.3 Classical methods for TSC

Over the years, different strategies have been proposed to address the TSC problem. A fixed-time signal control scheme has been widely used for managing traffic lights in urban areas. This strategy requires the determination of optimum TOD breakpoints for establishing TOD intervals, which are subsequently used for obtaining the predefined green splits for each split (green times) using Webster's formula or some other optimization tools [5]. However, the fixed-time signal control strategy is suitable for stable and nearly homogenous traffic patterns. Alternatively, studies have focused on actuated and traffic responsive TSC schemes for dynamic traffic control and management. In such traffic control schemes, signal cycle length and green splits are adjusted according to real-time traffic data collected from sensors installed on each approach. Though actuated TSC strategies overcome some limitations of the former methods, they do not work well under all traffic and adverse conditions. TSC problem was initially addressed using various probability and regression-based methods [6, 7]. However, for oversaturated and undersaturated traffic conditions, such methods do not provide reliable solutions. Few notable classic TSC strategies proposed during the last few decades include: SCOOT [8], SCAT [9], MAXBAND [10], CRONOS, PRODYN [11], TRANSYT [12], RHODES [13], OPAC [14], and FUZZY LOGIC [15]. Few other methods recently used for traffic light setting are ARRB [16], TRRL [8], and HCM [17]. In addition, to signal control strategies, traffic light design could be isolated intersection based or coordinated. Isolated intersections signal schemes have limited benefits compared to coordinated strategies that consider the network of intersections.

1.4 Limitations of classical TSC strategies

The timing of traffic signals significantly influences the performance of the transportation system. Obtaining the optimal signal timing plan for a network in its entirety is challenging due to the stochastic and non-linear characteristics of the traffic system. From a computational perspective, the signal control optimization problem under the influence of several constraints is a highly non-linear and non-convex problem. To reduce the complexity of problem, studies have assumed partial convexification for obtaining the optimal signal plans [18, 19]. It has been shown that traffic light optimization belongs to the family of NP-complete problems whose complexity increases dramatically for real-world and more extensive transportation networks with prolonged study periods. Classical optimization methods used in this regard are not suitable for a variety of reasons. For example, they are sensitive to initial estimates of solution vector and require gradient computation of constraints and the objective functions. Further, the discrete nature of signal timing plan and phasing sequence limit the application of traditional optimization approaches. Similarly, classical signal control optimization techniques are usually more suited to isolated intersections. They are not scalable for large urban transport networks where the interdependence of traffic signals across multiple intersections may be explored. Hence, such methods do not consider the interdependencies and connectivity of traffic signals vital for large-scale urban transport networks.

1.5 Metaheuristics for TSC: the new frontier

Metaheuristics techniques, including and swarm intelligence and evolutionary algorithms, have emerged as appealing alternatives to classical optimization methods for addressing signal control problems. They can be easily adapted for solving signal optimization problems with mixed types of continuous and discrete variables on large-scale transportation systems. Metaheuristics are based on approximate random methods and involve an iterative master process that can efficiently provide high-quality, acceptable solutions with relatively low computational efforts [20]. No prior information regarding the search space characteristics is required. In addition, metaheuristics do not rely on gradient information of the objective functions and the associated constraints with reference to signal timing variables. Further, the process of finding the optimal solution is simple and straightforward. Entailing less complexity than exact methods means that metaheuristics could be easily implemented to solve non-linear complex optimization problems. Furthermore, for many large-scale engineering problems that involve uncertainties (such as traffic flow), obtaining near-optimal solutions within a reasonable time is acceptable. Owing to these benefits, several metaheuristics techniques have been successfully applied for solving TSC optimization problems. Metaheuristics aim at obtaining the optimal values/ranges for various signal parameters that influence the performance of signalized intersections and include variables such as cycle length, green splits, phase sequence, offsets, change interval, etc. These parameters of interest are also known as decision variables. Constraints conditions for signal optimization include lower and upper cycle length, green splits thresholds, etc.

Metaheuristics have been widely applied to solve the TSC problems under a single objective framework known as mono-objective optimization. The single objective optimization can be classified into four main types: i) travel time minimization, ii) delay minimization, iii) throughput maximization, and iv) fuel consumption and exhaust emissions (CO, CO_2, NO_x, HC_s) minimization. Mono-objective optimization of traffic signals has some benefits; however, field traffic is highly complex, non-linear, and stochastic in nature, and quite often, the application of multi-objective optimization becomes inevitable. In the process of finding the optimal signal control parameters, traffic engineers usually deal with multiple conflicting objectives. They are seldom interested in knowing the single-objective-based best solution without considering the other objectives. It is quite possible that an indented improvement in one of the objectives may lead to the deterioration of others. Therefore, it is essential to obtain a reasonable trade-off among various clashing objectives while optimizing the signal timing parameters. To address this issue, researchers have proposed bi-objective or multi-objective metaheuristic frameworks which involve more than one objective function to be optimized concurrently. Adoption of multi-criteria/objectives metaheuristics for signal optimization is rational as well as more beneficial.

1.6 Study objectives

This study provides a comprehensive review of metaheuristics techniques applied to signal control optimization. The surveyed literature is categorized based on the types of metaheuristics used, i.e., evolutionary algorithms and swarm intelligence techniques. A total of over 15 metaheuristics optimization techniques in traffic signal control and optimization are presented. Literature is summarized based on classification of techniques, considered optimization objectives, decision variables, and constraints conditions. Finally, based on the identified literature gaps, major challenges and prospects for future research are also proposed.

1.7 Paper organization

The remainder of this work is organized as follows. Section 2 provides research methods and publication analysis of signal control optimization using metaheuristics. Section 3 reviews evolutionary algorithms' metaheuristics for signal optimization. Section 4 provides a summary of swarm intelligence techniques in the context of the subject domain. Section 5 and 6 presents surveys of trajectory-based metaheuristics and few others for TSC optimization. Finally, Section 7 presents the review conclusions and outlines the current challenges and recommendations for future research.

2. Methodology

The relevant literature on TSC was searched (in May 2021) using a detailed systematic review (SR). SR is a formal and standard protocol for performing a review study. To ensure that findings were reached in a valid and reliable manner, the study adopted a three-staged approach, i.e., i) planning, ii) execution, and iii) analysis. The planning stage involved defining the research scope and aims, setting the inclusion and exclusion criteria, and developing the review protocols. The execution stage involved a systematic search using relevant search strings. The relevant publications were meticulously selected by browsing through different electronic databases such as "Google Scholar," Science Direct," Wiley Online Library," "Scopus," "Web of Science," and "IEEE Xplore." To explore these databases, the following "Keywords" were used: "signalized intersections," "traffic congestion," "traffic signal control," "traffic signal timing optimization," "traffic control through metaheuristics," "intelligent traffic control," "dynamic traffic manage-ment," "traffic simulation and optimization," "multi-objective traffic control," etc. Titles, keywords, and abstracts of all the downloaded documents were reviewed to determine the appropriate selection of articles for the current study. Additional appropriate publications were added to the list by looking at the references selected

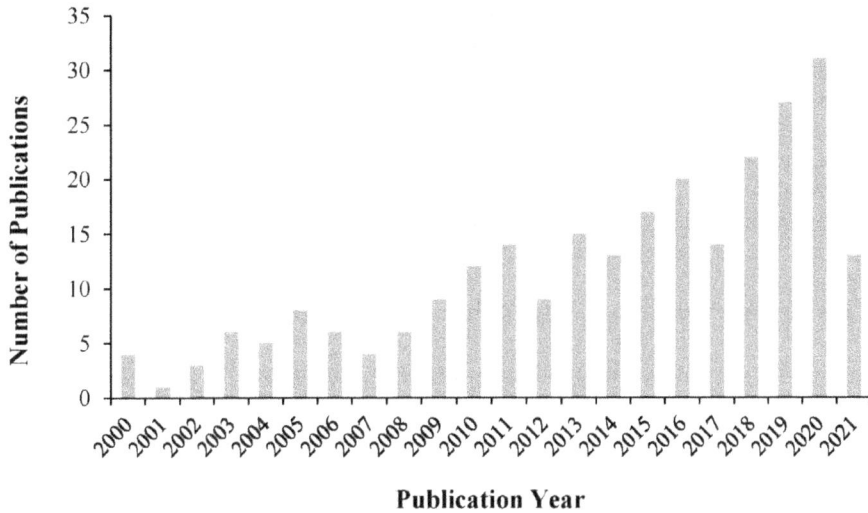

Figure 1.
Chronological distribution of indexed publications on traffic signal optimization using swarm intelligence and evolutionary computation techniques (period 2000–2021).

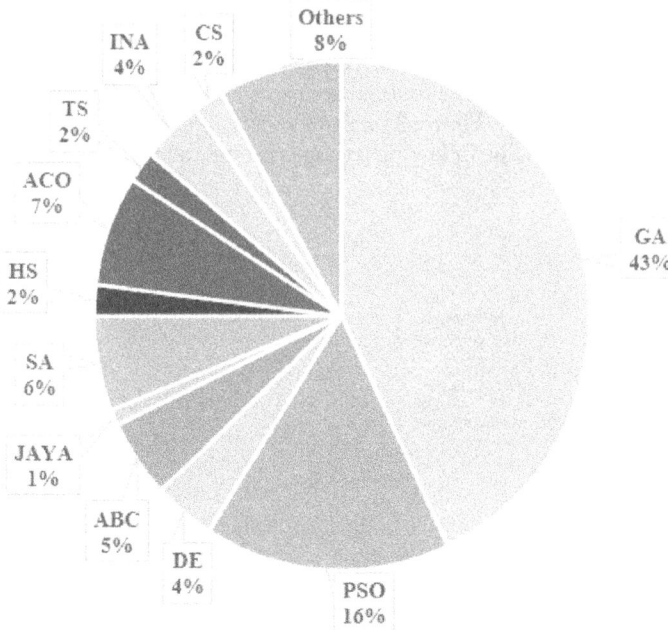

Figure 2.
Percentage distribution of indexed publications on traffic signal optimization based on metaheuristic type.

publications. Publications were searched irrespective of publication year and the number of citations to have the maximum number for initial consideration. Duplicate articles found in various databases were also identified and removed. Non-academic publications, such as magazine articles, company reports, newspapers, presentations, and interview transcripts, were excluded. Finally, the analysis stage involved the classification, categorization, and summarization of the main theme of selected articles.

Figure 1 presents the chronological distributions of shortlisted publications in which metaheuristics are used for solving traffic signal control optimization. It may be observed from the publications reporting in **Figure 1** that is there is a growing trend in the application of metaheuristics in the subject domain. **Figure 2** shows the percentage distribution of published studies in the area of traffic control optimization based on the type of metaheuristic applied. It may be observed from the Figure that the Genetic Algorithm (GA), Particle Swarm Optimization (PSO), and Ant Colony Optimization (ACO) have been widely used for signal optimization.

3. Review of evolutionary algorithms (EAs) for TSC

This section reviews the previous studies in the literature that applied evolutionary algorithms (EAs) for traffic signal control and optimization. EAs are the most widely used metaheuristics optimization techniques across diverse fields of science and engineering. EAs are population-based random search techniques and are inspired by Darwin's theory of natural theory of evolution. The EAs contain a population of individuals, each symbolizing a search point in the feasible solution space exposed to a common learning process while proceeding among different generations. EAs begins with the initialization of random population, which are then subjected to selection, crossover, mutation through various generations so that offsprings generated evolve toward more favorable regions in the search space. At each generation, the fitness of the population is evaluated, and those with better fitness values are selected and recombined that have an increased probability of improved fitness. The program is iteratively repeated until it converges to the best (or near-optimal) solutions. The basic structure of EAs remains similar for all the algorithms under its family. **Figure 3** presents the sample structure of EAs and their working principle. The following passages provide a brief explanation of

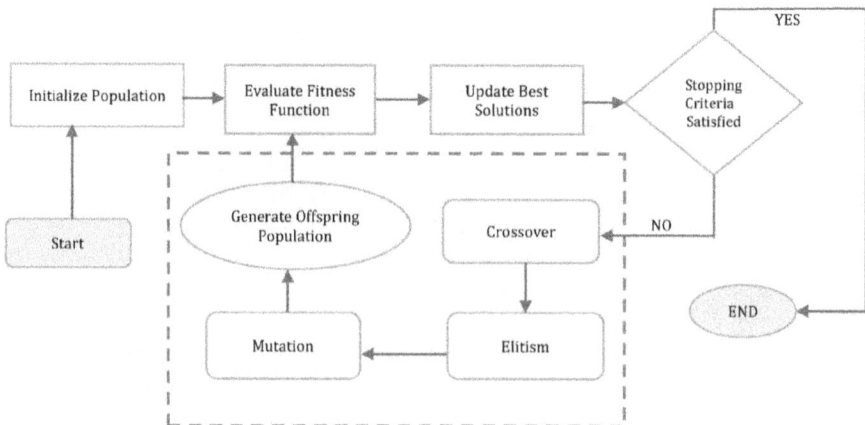

Figure 3.
General flow depicting the search mechanism of EAs.

S.No	Metaheuristic Used	Optimization Objectives							Reference
		Delay	Stops	throughput	Travel time	Queue	Emissions	Fuel Consumption	
1	GA	✓							[21]
2	GA	✓	✓						[22]
3	DE						✓	✓	[23]
4	GA	✓							[24]
5	DE	✓		✓					[25]
6	DE	✓							[26]
7	GA	✓					✓		[27]
8	GA	✓					✓		[28]
9	GA and DE	✓							[29]
10	GA	✓					✓	✓	[30]
11	DE	✓				✓			[31]
12	GA						✓	✓	[32]
13	GA	✓							[33]
14	NSGA			✓		✓			[34]
15	NSGA-II	✓	✓				✓	✓	[35]
16	GA	✓							[36]
17	DE	✓		✓					[37]
18	GP	✓							[38]

Table 1.
Summary of previous studies on traffic signal optimization using EAs.

various EAs employed in the field of traffic signal optimization. **Table 1** presents a summary of previous studies that have applied EAs for traffic signal control and optimization.

3.1 Genetic algorithm

Genetic algorithm is the most widely used method for traffic light optimization. John Holland initially proposed the GA metaheuristic in 1975 [39]. GA search mechanism for finding the optimal solution of an objective function mimics the natural selection process of the evolutionary theory of nature, which supports the "survival of the fittest" concept. It is a population-based technique that involves the ranking of individual members of the population according to their fitness.

The search process of the optimal solution begins with the initialization of a random population of solutions. The offspring population is created by iteratively applying various genetic operators such as crossover, mutation, elitism, etc. until the stopping criteria are satisfied. In literature, many studies have demonstrated the robustness of GA for adaptive traffic signal control. For example, Foy et al. utilized GA for traffic light optimization, considering delay time minimization as the objective function [36]. The number of initial GA generations was varied over five GA traffic runs. The optimal fitness value was achieved for populations ranging between the 20th to 30th generations with an average vehicle waiting time of around 40 seconds. GA was noted to yield rational signal timing plans reducing the timing delay significantly compared to the existing traffic control scheme. In their study, Rahbari et al., studied that traffic control at the signalized intersection with GA could reduce the congestion [40]. Yang and Luo adopted a hybrid GA simulated annealing (GA-SA) for signal control optimization at isolated signalized intersections considering delay as the objective function [41]. Empirical results showed that GA produced a rational signal timing plan compared to fixed control scenarios. A study conducted by Mingwei et al. proposed the application of multi-objective for intelligent traffic management at an isolated signalized intersection for a case study in China [42]. The considered optimization objectives included; average vehicle delay, vehicular stops, and fuel consumption. It was found that the optimized signal timing plan from GA significantly improved the considered traffic performance measures.

In another study, Turki et al. proposed a multi-objective NSGA-II to optimize various measures of effectiveness (such as delay, stops, fuel consumption, and emissions) at isolated signalized intersections in the city of Dhahran, Saudi Arabia [35]. Study results were compared with Synchro traffic simulation and optimization tool, and the results for a typical intersection are shown in **Figures 4** and **5**. All the performance measures witnessed considerable improvement for the optimized signal timing plan obtained using an NSGA-II optimizer. **Figure 4 (a–d)** depicts the evolution of the four selected performance measures (delay, stops, fuel consumption, and emissions) against the number of iterations for three random initial populations. It may be noted that all the algorithms converged to their respective objective functions at approximately 70 to 100 generations. Comparing the random initial populations, population size 30 for all cases yielded the best results.

Figure 5 shows the performance comparison of NSGA-II and Synchro signal control strategies for the selected measures of effectiveness (delay, stops, fuel consumption, and emissions). It may be noted from the Figure that the NSGA-II optimizer outperformed the Synchro results for all the performance measures.

Li et al. also investigated the applicability of NSGA-II for solving signal control optimization problems [34]. Average queue ratio and vehicle throughput were the objective functions. The algorithm's results were validated on a microscopic traffic

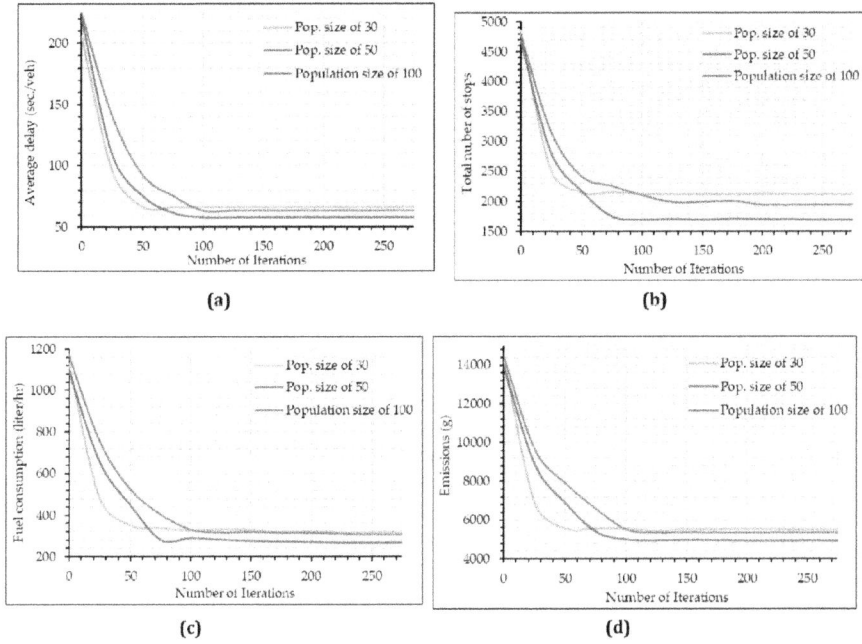

Figure 4.
Evolution of different performance measures against NSGA-II iterations; (a) delay evolution, (b) number of vehicle stops evolution, (c) fuel consumption evolution, (d) emission evolution. Reprinted with permission from Ref. [35] copyright (2021), MDPI.

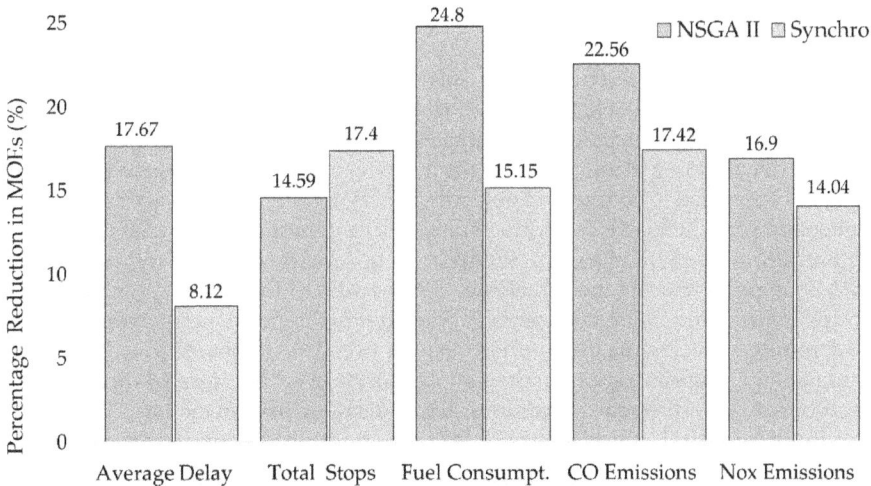

Figure 5.
Comparison of NSGA-II and synchro optimizers for various traffic performance measures. Reprinted with permission from Ref. [35] copyright (2021), MDPI.

simulation tool, VISSIM. Kwak et al. developed a GA traffic optimizer to evaluate the influence of traffic light setting on vehicle fuel consumption and emissions [32]. Model results were compared with TRANSIM, a microscopic traffic simulator. It was observed that the proposed GA traffic optimizer could reduce exhaust emissions by approximately 20% and fuel consumption in the range of 8–20%. In another study, Kou et al. employed multi-criteria GA for optimizing the signal timing plan of signalized junctions and compared the results with the highway capacity manual (HCM) method [28]. The study considered several optimization objectives such as stops, delays, and emissions. A reasonable trade-off established an optimal Pareto front among different conflicting objectives. Study results demonstrated the superior performance of the proposed GA traffic control scheme compared to the HCM method in terms of all the optimization objectives. Guo et al. developed a model for area-wide intersection traffic control in the central business district (CBD) area of Nanjing, China [43]. Capacity ratio, turning movement delay, and travel time was the three chosen objective functions. Computational experiments results showed significant mobility improvement compared to existing conditions. Study results were also validated in PARAMICS traffic simulation tool. In their study, Dezani et al. have shown that simultaneous optimization of traffic lights via GA and vehicle routes could significantly reduce the vehicle travel time compared to optimization considering only routes [44]. In another study, Tan et al. proposed a new Decentralized Genetic Algorithm (DGA) for signal timing optimization of traffic networks under oversaturated traffic conditions [45]. Average vehicle delay was used as the performance metric to evalauate the performance of proposed algorithm. Simulation results showed that DGA could effectively optimize the traffic light setting and reduced the average network delay.

3.2 Differential evolution (DE)

Differential evolution is another population-based metaheuristic technique initially proposed by K.V. Pricein 1995 [46]. DE is characterized by its robustness, fast convergence to the objective function, and simplicity. Though the method has been successfully used for numerous applications across different disciplines, only a few studies have adopted DE for urban traffic control and management [25–29]. For example, in their recent study, Jamal et al. compared the performance of GA and DE for optimizing traffic lights at isolated signalized intersections in the city of Dhahran, Saudi Arabia [29]. Average delay time minimization was the objective function. The study concluded that both GA and DE could yield intelligent and rational signal timing plans, reducing the intersection average delay between 15 and 35%. DE was noted to converge to objective function faster than DA over several simulation runs. Similarly, in another study, Liu et al. proposed bacterial foraging optimization-based DE algorithm for optimizing delay at signalized intersections [37]. To improve convergence precision, DE was utilized for updating the bacteria position during the chemotaxis process. The proposed scheme yielded very promising results, reducing the intersection delay by over 28% compared to only 5% obtained by GA optimization. In their study, Korkmaz et al. suggested three different types of delay differential evolution-based delay estimation models (DEDEM), i.e., linear, quadratic, and exponential [47]. The researchers reported that all the proposed models effectively predicted the vehicle delay estimates in terms of relative errors between estimated and simulated values; however, quadratic DEDEM methods outperformed other models. Ceylan also approached the signal control optimization problem using the metaheuristic DE and Harmony-Search (HS) for network-wide traffic control and optimization [48]. Study results showed that DE algorithms yielded better results compared to HS.

In another research study, Yunrui et al. proposed multi-agent fuzzy logic control based on DE to optimize delay and queue length through a network of eleven intersections in the urban traffic context [31]. DE was used to decide and optimize the parameters of the fuzzy systems because it is easy to understand and implement.

Empirical results revealed that the proposed method could substantially improve the network performance measures such as average vehicle delay, traffic throughput, and queue length. In a recent study, Liu et al. have proposed an improved adaptive differential evolution (ADE)-based evolvable traffic signal control (EvoTSC) scheme for global optimization of different traffic performance measures on large scale urban transportation networks [49]. The proposed TSC method was compared with a conventional TSC scheme on two practical and three synthetic transportation networks with varying traffic flow demands and different physical scales. Comparison results indicated that the DE-based EvoTSC method significantly outperformed its counterpart under all the considered scenarios. Zhang et al. also applied an online intelligent urban traffic signal control approach using multi-objective DE for real-time traffic control and optimization [50]. Experimental results showed that the proposed approach provides a more robust configuration of traffic signal phases and has relatively better real-time performance than the traditional traffic control scheme.

3.3 Genetic programming (GP)

Genetic programming (GP) is another population-based metaheuristic technique that belongs to the family of evolutionary algorithms [51]. GP is an extension of GAs that allows for deep exploration of space on computer programs. GP starts with a population of random programs (candidate solutions) that are fit for applying evolutionary operators similar to genetic processes, thereby simulating the fundamental principles of Darwin's evolution theory [52]. GP follows an iterative process to breed the solutions to problems using the probabilistic selection procedure for the carryover of fittest solutions to the offerings by applying genetic operators such as crossover and mutation. In literature, not many studies have focused on applications of GP for traffic analysis and management in urban transport networks. Montana and Czerwinski used a hybrid GA with strongly typed GP (STGP) for intelligent control and optimization of evolving traffic signals on a small-scale transport network [53]. Numerical simulation results showed that the proposed hybrid STGP model could effectively improve network performance under varying traffic demands.

A study conducted by González also proposed the application of GP for solving signal control problems [54]. This study considered four different traffic scenarios with properties and traffic conditions in a previous study [55]. Study results were also validated using the microscopic traffic simulator tool SUMO. Findings showed that GP could provide competitive and robust results for all the tested scenarios. However, the highway/network scenario had a more pronounced performance improvement (having an improvement of 10.34%) than the isolated intersection scenario (with an improvement of 4.24%). In another study, Ricalde and Banzhaf adopted an improved GP with epigenetic modifications for traffic light scheduling and optimization under dynamic traffic conditions [56]. Extensive simulation analysis revealed that the proposed model improved the network performance compared to standard GP and other previously used methods. This study, however, did not use any real-world data for validation purposes. In another study, the authors used a similar GP approach with epigenetic modifications (EpiGP) to design and evolve traffic signals using real-time field traffic data [38]. Results indicated significant improvement in network performance compared to conventional methods, including standard GP, static, and actuated traffic control techniques.. Over 12% improvement in average delay was reported under high-density traffic conditions.

4. Review of swarm intelligence (SI) techniques for TSC

This section reviews the previous studies in the literature that applied swarm intelligence (SIs) techniques for traffic signal control and optimization. SI is another class metaheuristics that are increasingly used for various engineering and industrial applications. The search mechanisms of SI are believed to be inspired by human cognition representing the individual's interaction in a social environment. For this reason, SI techniques are also sometimes called "behaviorally inspired algorithms." In SI algorithms, each swarm member has a stochastic behavior due to their perception of the neighborhood and acts without supervision. By collective group intelligence, swarm utilizes their resources and environment effectively. The primary attribute of a swarm system is self-organization, which assists in evolving and obtaining the desired global level response by effective interactions at the local level. Just like EAs, SIs are population-based iterative procedures. After randomly initializing the population, individuals are evolved across different generations by mimicking the social behavior of animals or insects to reach the optimal solutions. However, SIs do not involve the use of evolutionary operators like crossover and mutation like EAs. Instead, a potential solution modifies itself based on its relationship with the environment and other individuals in the population as it flies through the search space. The following passages provide a brief explanation of various swarm intelligence techniques employed for solving signal control optimization problems. **Table 2** presents a summary of previous studies that have applied SIs for traffic signal control and optimization.

4.1 Particle swarm optimization (PSO)

Particle swarm optimization is a population-based swarm intelligence technique that was first introduced in 1995 by Eberhart and Kennedy. In the PSO algorithm, every potential solution is referred to as a particle representing a location in the problem space. The entire population of potential solutions (particles) is called the swarm. PSO search mechanism for global optima is inspired by birds in which each particle can update its velocity and position by using local and global best values. PSO is yet another widely used optimization algorithm for signal control problems. For example, Celtek applied PSO for real-time traffic control and management in the city of Kilis city in Turkey [77]. Algorithm performance was investigated in real-time using the SUMO traffic simulator. Social Learning-PSO was introduced as an optimizer for the traffic light. Empirical results obtained using the proposed PSO architecture resulted in travel time by 28%. The algorithms performed well both for undersaturated and oversaturated traffic conditions. Gokcxe and Isxık proposed a microscopic traffic simulator VISSIM-based PSO model for optimizing vehicle delay and traffic throughput using field data from28 signalized roundabout in Izmir, Turkey [64]. The simulation tool was used to evaluate the solutions obtained by PSO. Optimization of traffic signal head reduced the average delay time per vehicle by approximately 56% and increased the number of passing vehicles by 9.3%. In their study, Jia et al. employed multi-objective optimization of TSC using PSO [72]. The optimization objectives included average vehicle delay, traffic capacity, and vehicle exhaust emissions. The validity of the algorithm was examined by applying it to the real-time signal timing problem. The suggested algorithm provided competitive performance for all the MOEs compared to other efficient algorithms such as NSGA-II, IPSO, and GADST.

Abushehab et al. compared PSO and GA techniques for signal control optimization on a network of 13 traffic lights [78]. SUMO was used as a simulator tool for the network. Both the algorithms yielded systematic and rational signal timing plans;

S.No	Metaheuristic Used	Optimization Objectives							Reference
		Delay	Stops	throughput	Travel time	Queue	Emissions	Fuel Consumption	
1	ACO	✓	✓	✓					[57]
2	AIS			✓		✓			[58]
3	GWO					✓			[59]
4	ABC	✓	✓						[60]
5	ACO	✓	✓						[61]
6	BA				✓	✓			[62]
7	CS	✓							[63]
8	PSO	✓			✓				[64]
9	PSO	✓	✓						[65]
10	BA				✓				[66]
11	PSO	✓							[33]
12	PSO							✓	[67]
13	ABC	✓					✓		[68]
14	ABC	✓	✓						[60]
15	PSO	✓				✓	✓		[40]
16	ACO	✓							[69]
17	CS	✓							[70]
18	ACO			✓	✓				[71]
19	PSO	✓		✓			✓		[72]

S.No	Metaheuristic Used	Optimization Objectives							Reference
		Delay	Stops	throughput	Travel time	Queue	Emissions	Fuel Consumption	
20	PSO					✓			[73]
21	PSO	✓							[74]
22	INA	✓		✓					[75]
23	FFA					✓			[76]

Table 2.
Summary of previous studies on traffic signal optimization using SI techniques.

however, some algorithm variants were found to be more efficient than the others. In their study, Angraeni et al. proposed a modified PSO (MSPO) and fuzzy neural network (FNN) for optimizing signal cycle length at an isolated intersection [79]. Simulation results using PSO led to a reduction in MSE value from 6.3299 to 2.065, while network performance was improved by 4.26%. The accuracy of the training process using MPSO was higher than FNN. Chuo et al. reported a significant decrease in vehicle queue length by using PSO as a traffic signal optimizer [73]. In another study, Garcıa-Nieto et al. applied PSO to optimize the cycle program of 126 traffic signals located in two large and heterogenous metropolitans of cities of Bahıa Blanca in Argentina and Malaga in Spain [80]. The Obtained solutions were validated using the traffic simulation package SUMO.

In comparison to the existing pre-defined traffic control schemes, PSO achieved significant quantitative improvement for both the objectives, i.e., overall journey time (74% improvement) and the number of vehicles reaching their destinations (31.66%) improvement). In another study, a researcher proposed an improved PSO architecture by combining traditional PSO with GA for multi-objective traffic light optimization. The selected performance indexes included vehicular emissions, vehicle delay, and queue length [40]. The authors reported that the improved PSO method has a quick response and higher self-organization ability which is beneficial for improving the efficiency of traffic signal control.

Olivera et al. investigated the applicability of PSO to reduce vehicular exhaust emissions (CO and NOx) and fuel consumption considering large-scale heterogeneous urban scenarios in the cities of Seville and Malaga in Spain [67]. Study results showed that the proposed signal control strategy could significantly reduce the exhaust emission (CO by 3.3% and NOx by29.3%) compared and fuel consumption (by 18.2%) compared to signals designed by human experts. In their study, Qian et al. designed a simulation protocol for traffic different signal parameters such as cycle, green signal ratio, and phase difference using three Swarms Cooperative-PSO algorithms [74]. The considered optimization objectives included average vehicle delay and average parking number per vehicle. Algorithm simulation results were validated using traffic simulator CORSIM. Lo and Tung compared the performance of PSO and GA-based signal control along four intersections on an urban arterial and noted that the PSO algorithm outperformed GA both in terms of speed convergence and accuracy of search [81]. A couple of other recent studies also demonstrated the adequacy and robust performance of PSO for TSC and optimization [82, 83].

4.2 Ant colony optimization (ACO)

Ant Colony optimization is a swarm intelligence method-based optimization technique that mimics the natural behavior of ants in finding the shortest path from an origin to a food source [84]. In ACO, the path of every ant from origin to destination is considered as a possible solution. ACO has been widely used for traffic signal optimization. In their study, Putha et al. used ACO for traffic signal coordination and optimization in the context of an oversaturated urban transport network [85]. The authors reported that ACO could provide reliable solutions of optimal signal timing plan compared to GA. Yu et al. also applied ACO for intelligent traffic control at signalized intersections considering vehicle waiting time as the optimization objective [86]. The authors reported that ACO outperformed the traditional traffic actuated scheme, predominantly during traffic flow periods. He and Hou also proposed the application of a multi-objective ACO algorithm for the timing optimization of traffic signals [57]. Several parameters such as vehicle delay, number of stops, and traffic capacity performance indices were chosen as performance indexes. Numerical simulation results demonstrated that ACO is a simple and robust technique for signal control optimization problems. The proposed ACO technique significantly improved the selected performance indicators compared to Webstar and GA algorithms.

In another study, ACO optimized the timing plan for traffic lights at isolated signalized intersections [61]. All the selected intersection measures of effectiveness (MOEs), including vehicle delay, parking rate, and the number of stops, were improved by a fair margin. Sankar and Chandra proposed a multi-agent ACO for effective traffic management on a network level [69]. The authors concluded that the method could be pretty useful in reducing average vehicle delays and traffic congestion under varying traffic conditions. Haldenbilen et al. developed an ACO-based TRANSYT (ACOTRANS) model for area traffic control (ATC) through a coordinated signalized intersection networks under different traffic demands [87]. A total of 23 links were considered for the analysis, and the network Disutility Index (DI) was chosen as the primary performance index. A comparative analysis of the network's PI obtained using TRANSYT-7F with hill-climbing (HC) optimization and TRANSYT-7F with GA was also performed. Study results showed that the proposed ACOTRANS improved the network's PI by 13.9% and 11.7% compared to its counterparts TRANSYT-7F optimization with HC and GA. Li et al. compared ACO and Fuzzy Logic for optimizing traffic signal timing in a simulated environment [88]. Traffic capacity and vehicular delay were considered as the objective functions and did not consider pedestrian traffic. The validity of proposed algorithms was tested using actual time-period and conventional algorithms. Jabbarpour et al. conducted a detailed review of the literature focused on applying ACO evolutionary algorithms for the optimization of vehicular traffic systems [89].

Rida et al. proposed ACO for real-time traffic light optimization problems at isolated signalized intersections [71]. Objective functions include minimizing the vehicle waiting time and increasing the traffic flow. The proposed model yielded robust performance compared to fixed time signal controller and other dynamic signal control strategies. Renfrew and Yu, in their studies, also reported that ACO demonstrated robust performance compared to actuated control in optimizing signal timing plan, particularly under high traffic demand [90, 91]. Srivastava and Sahana proposed a novel hybrid nested ACO model intending to reduce the vehicle waiting time at signalized intersections [92]. The proposed model was also compared with the hybrid nested GA model. Results showed that nested hybrid models outperformed traditional ACO and GA-based traffic control.

4.3 Artificial bee colony (ABC)

The traditional algorithms used for training carry some drawbacks of getting stuck in computational complexity and local minima. The artificial bee colony (ABC) algorithm is a revolutionary approach developed by Karaboga et al. [93]. ABC has good exploration capabilities in finding optimal weights during the training process [94]. ABC algorithm operates on the principle of foraging behavior of honeybees in seeking quality food. Each cycle of the search comprising three steps: sending employed bees onto the food source to measure nectar amount; selecting food source by onlookers once the information is shared by employed bees, and sending the scouts for discovering new food source [95].

ABC algorithm is widely used in optimizing traffic-related problems by previous researchers [60, 68, 96]. Zhao et al. investigated a typical intersection as a case study at Lanzhou city [60]. The green time length of each phase of the signal cycle and signal cycle were considered as decision variables. Favorable convergence was achieved using different setting parameters of the algorithm. The effect of signal cycle on control targets resulted that vehicle delays will increase with the signal cycle; however, the stops will decrease. In comparison to non-dominating sorting genetic algorithm and webster timing algorithm, ABC manifested better convergence. In another study, Dell'Orco et al. developed TRANSYT-7F to investigate network performance index (PI) for optimizing signal timing [96]. Results revealed that PI's of the network in the case of ABC improved by 2.4 and 2.7% compared to genetic algorithm and hill-climbing method.

4.4 Cuckoo search (CS)

Cuckoo search (CS) is a recently developed metaheuristic algorithm developed by Yang and Deb [97], inspired by the natural breed parasitism of the cuckoo species. For understanding its working principle, consider that each bird lays one egg at a time and dumps it in a random nest which represents a single solution. The nest with high-quality eggs will be moved to the next generation. The number of host nests is fixed, and the egg laid by the cuckoo is discovered by the host bird. In this situation, the host bird either gets rid of the egg or abandons the nest by developing a new nest [98]. Few studies interpret CS as more efficient than PSO and GA [97].

Araghi et al. employed neural networks (NN) and adaptive neuro-fuzzy inference system (ANFIS) to optimize the results of CS in the case of intelligent traffic control [63]. The results were compared to that of the fixed time controller. It was revealed that the CS-NN and SC-ANFIS showed 44% and 39% improved performance against the fixed-time controller. Similarly, in another study, the authors evaluated the performance of ANFIS using CS for optimization of controlling traffic signals for an isolated intersection [70]. Improved performance of ANFIS-CS was obtained against fixed-time controller.

4.5 Bat algorithm (BA)

Bat algorithm (BA), initially developed by Xin-she yang in 2010, is inspired by the echolocation of microbats [99]. The working principle of BA encompasses three basic steps: bats use echolocation to sense the distance bifurcating the food and barrier; bats randomly fly with variable loudness and wavelength.; bats automatically adjust their wavelength and pulse depending upon the proximity of food/prey [100].

Srivastava, Sahana used BA to determine the wait time at a traffic signal for the discrete microscopic model [66]. The study was based on 12 nodes and four intersections. The results were compared to GA. Relatively higher performance was obtained for BA algorithm as compared to GA. Jintamuttha et al. carried experimental simulation for the green time of intersection for ten cycles per run [62]. The results of the experiment were optimized using BA. The average queue length and waiting time improved due to optimization.

4.6 Artificial immune system (AIS)/immune network algorithm (INA)

The immune network algorithm (INA) or artificial immune system (AIS) is another useful optimization algorithm recently practiced for signal control optimization problems. As its name suggests, the working mechanism of this algorithm is inspired by the biological immune system. Immune cells have receptors that can detect harmful pathogens and activate antibodies to fight them, leading to their elimination [101]. Louati et al. applied INA to optimize queue, delay, and traffic throughput at signalized intersections under varying traffic demands [75]. It was found that INA outperformed traditional fixed-time adaptive traffic control strategies and validated the study results through VISSIM, a microscopic traffic simulation platform. In another study, Trabelsi et al. evaluated the performance of AIS to detect and rationally control anomalous traffic conditions through a network of signalized intersections [58]. Simulation results proved the adequacy and robustness of the proposed AIS-based signal control method.

Darmoul et al. employed multi-agent immune network (INAMAS) for optimal control and management of interrupted traffic flow at signalized intersections [102]. The proposed INAMAS models offered an intelligent mechanism that could explicitly capture the disturbance-related knowledge of traffic fluctuations. To demonstrate the efficacy of the proposed model, the authors compared its performance against two widely used signal control strategies, namely fixed-time control and LQF-MWM (longest queue first –maximal weight matching) algorithm. The suggested INAMAS scheme provided a competitive performance in terms of chosen performance indicators, i.e., vehicle queue and waiting times under extreme traffic conditions involving high traffic volume and block approaches. **Figure 6a** plots the average vehicle delay for all the three signal control strategies under various traffic scenarios [102]. For scenario 1 (moderate traffic congestion), the INAMAS algorithm produces approximately a 24% reduction in average delay values compared to the LQF-MWM strategy. For scenario 2 (high-density traffic), the proposed INAMAS optimizer decreased the average delay by nearly 32%. For scenario 3 (extreme congestion), the corresponding improvement by the INAMAS algorithm is about 28%. **Figure 6b** depicts the relationship between the total network delay and simulation time (in minutes) for all three signal optimization strategies [102]. It is evident from the results in **Figure 6b** that during the first 5 minutes, all the controllers have comparable performance. At the end of simulation analysis (after 5 hours), when the traffic density reaches 9600 vehicles per hour, the INAMAS controller achieved better performance compared to others, showing its superior capability to manage large and complex traffic networks.

Moalla et al., in their study, also demonstrated the robustness of AIS for controlling traffic at isolated signalized intersections [103]. However, the authors also emphasized that validation of the proposed AIS scheme is challenging and should be handled carefully. In another study, the author highlighted AIS-based traffic control's significance for network-wide traffic management [104]. Comparative results with TRANSYT 7F showed the superior performance of AIS approach. Galvan-Correa et al. proposed a new metaheuristic known as the micro artificial

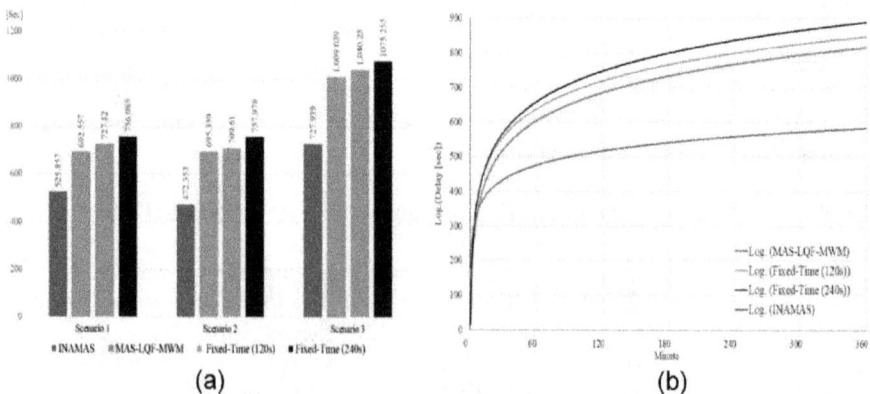

(a)   (b)

Figure 6.
(a) Comparison of average total delay per vehicle from various optimizers (b) cumulative network delay for scenario 1 for various optimizers Ref. [102].

immune systems (MAIS) to optimize vehicular emission and traffic flow in the city of Mexico [105]. The performance of the suggested MAIS technique was compared with several other metaheuristics, including GA, DE, SA, PSO. Results showed that MAIS achieved better results compared to most of the other metaheuristics. In a recent study, Qiao et al. proposed a novel hybrid algorithm, known as the Immune-Fireworks algorithm (IM-FWA) for effective traffic management on large-scale urban transportation networks [106]. The proposed hybrid algorithm was developed based on fireworks and artificial immune algorithms. A hierarchical strategy was proposed in the framework to avoid possible offsets conflicts and reasonable configuration of intersection offsets. Simulation results showed that the proposed IM-FWA could successfully overcome the shortcomings of FWA and AIS algorithms by providing a better and more rational signal timing plan to effectively reduce traffic flow delays.

4.7 Firefly algorithm (FA)

The characteristic behavior of fireflies is animated by Yang [107] into a nature-inspired meta-heuristic swarm intelligent method called Bat Algorithm. In BA, all fireflies are assumed unisex, and attractiveness is proportional to their brightness, which in turn depends on the distance. Thus, the brightness can be considered a cost function, which is maximized in optimization.

Kwiecień, Filipowicz [studied optimizing costs controlled by queue capacity, maximal wait, and servers [76]. It was deduced that the use of FA could maximize the value of the objective function, and FA converges toward the optimal solution very quickly. Goudarzi et al. [108] investigated traffic flow volume by a probabilistic neural network method called deep belief network (DBN). FA was used to optimize the learning parameters of DBN. As a result, the proposed model predicted the traffic flow with higher accuracy compared to traditional models.

4.8 Gray wolf optimizer (GWO)

Gray wolf optimizer (GWO) is a new metaheuristic technique recently proposed by Mirjalili in 2014 [109]. GWO is inspired by the social hierarchy and hunting behavior of gray wolves. In GWO optimization, the wolves represent a solution set of candidate solutions. The hunting cycle in the GWO commences with the acquisition of a random population of candidate solutions (wolves) followed by identifying optimal prey's locations using a cyclic process. GWO has several advantages compared with evolutionary approaches, easy programming and implementation, algorithm simplicity, no need for algorithm-specific parameters, and lower computational complexity [110]. In recent years, GWO has been increasingly used in diverse disciplines. However, studies on its applications in transportation and traffic engineering in general and traffic control and optimization in particular are very few.

Teng et al. were the first to use a hybrid gray wolf and grasshopper algorithm (GWGHA) algorithm for timing optimization of traffic lights [111]. The obtained solutions were simulated in a microscopic traffic simulator package SUMO. The performance of the proposed GWGHA hybrid algorithm was compared with other metaheuristics like GWO, GOA, PSO, and SPSO2011. Results indicated that the proposed hybrid algorithm provided better solutions than its counterparts because it utilizes the feature of GWO for accelerating the convergence speed while using GOA to diversify the population. In another recent study, Sabry and Kaittan proposed a novel hybrid algorithm consisting of gray wolf and fuzzy proportional-integral (GW-FPI) for active vehicle queue management in an urban context [59].

The proposed traffic controller was compared with PI through repeated MATLAB simulations. Study results indicated the stable and robust performance of the proposed hybrid controller for queue management in a dynamic transport network with varying traffic flow demands.

5. Review of trajectory-based metaheuristics for TSC

This section surveys the previous works that applied trajectory-based meta-heuristics techniques) for traffic signal control and optimization. As the name suggests, these algorithms form search trajectories in solution space and iteratively improve the single solution in its neighborhood. Their exploration process starts from a random initial solution generated by another algorithm. At each stage, the current solution is replaced by a better offspring population. Trajector-based metaheuristics are mainly characterized by their internal memory sorting the state of search, candidate solution generator, and selection policy for candidate movement through generations. **Table 3** summarizes the previous works that applied trajectory-based search metaheuristics, hybrid metaheuristics, and others for traffic signal control and optimization.

5.1 Tabu search for signal control optimization

Tabu Search (TS) is a metaheuristic introduced by Fred Glover in 1986 to overcome the local search (LS) problem of existing methods [123]. TS allows the LS heuristic to diversify the search for solution space outside the local optima [124]. One of the important features of TS is its memory function, which can restrict few search directions for a more detailed LS, thereby making it easier to avoid local optimum solutions. By combining the greedy concept and randomization, the TS algorithm could provide an efficient solution to many optimization problems. In literature, only a few studies have focused on the application of Tabu search for signal control optimization. Hu and Chen proposed traffic signal control based on a novel greedy randomized tabu search (GRTS) algorithm considering travel time as the primary optimization objective [118]. GRTS results were compared with a GA-based traffic control scheme using data from a real city network to demonstrate the benefits of the proposed method. Numerical simulation results revealed that over 25% reduction in travel time might be achieved under medium to high traffic demands. In another study, Karoonsoontawong and Woller applied reactive tabu search (RTS) for simultaneous solutions of traffic signal optimization and dynamic user equilibrium problems on two transport networks in a simulated environment [119]. Three different variants of RTS were investigated based on deterministic or probabilistic neighborhood definitions. The performance of all the RTS variants was evaluated using three criteria such as solution quality, CPU time, and convergence speed. Simulation results showed that the RTS approach could provide promising results in terms of improving the overall network performance.

In a recent study, Hao et al. proposed a hybrid tabu search-artificial bee colony (TS-ABC) algorithm for robust optimization of signal control parameters in undersaturated traffic conditions at isolated signalized intersections [68]. This study considered two performance indexes such as average delay and mean-square error of average delay. The proposed signal control optimizer was validated using field data from an intersection in the city of Zhangye, China. Numerical simulation results compared with GA showed that the proposed TS-ABC is better in reducing the traffic delay under varying and heterogeneous traffic conditions. Chentoufi and Ellaia also proposed a hybrid particle swarm and tabu search (PSO-TS) for adaptive

S.No	Metaheuristic Used	Optimization Objectives							Reference
		Delay	Stops	throughput	Travel time	Queue	Emissions	Fuel Consumption	
1	SA-GA	✓							[112]
2	IM-FWA	✓							[106]
3	ISA-GA						✓		[113]
4	SA	✓	✓						[114]
5	HS	✓							[115]
6	HS	✓	✓	✓					[116]
7	JAYA	✓							[117]
8	TS				✓				[118]
9	TS-ABC	✓							[68]
10	TS				✓				[119]
11	PSO-TS	✓							[120]
12	WCO	✓							[121]
13	GHW-GHA	✓		✓					[111]
14	JAYA	✓							[122]
15	GW-FPI					✓			[59]

Table 3.
Summary of previous studies on traffic signal optimization using trajectory-based metaheuristics, hybrid metaheuristics, and others.

traffic lights timing optimization on real-time isolated signalized intersections in the context of Moroccan cities [120]. The authors also highlighted the significance of integrating the proposed PSO-TS model and VISSIM to achieve optimum average delay estimates. Simulation results demonstrated the superior efficiency of the PSO-TS technique against the traditional static models under oversaturated traffic conditions.

5.2 Simulated annealing (SA)

Simulated Annealing (SA), developed by Kirkpatrick et al. is inspired by the statistical mechanics of annealing in solids [125]. For understanding, consider a change in temperature, which causes a change in energy and movement of particles in solids. There is a sequence of decreasing temperature in annealing until criteria are met [126].

Li, Schonfeld [112] reported traffic signal time optimization using metaheuristic capabilities of SA with GA. It was concluded that SA-GA models outperform in optimization compared to individual SA and GA models. Similar results were reported by Song et al. in evaluating the optimized model for reducing traffic emissions on arterial roads [113]. Oda et al. [114] employed SA to optimize traffic signal timing and reported its improved performance as compared to traditional models.

6. Other metaheuristics for TSC

This section reviews the previous works that applied some other metaheuristics for traffic signal control and optimization. These include the harmony search algorithm, water cycle algorithm, and Jaya algorithm. **Table 3** summarizes the previous works that applied trajectory-based search metaheuristics, hybrid metaheuristics, and others for traffic signal control and optimization.

6.1 Harmony search (HS)

The metaheuristic harmony search (HS) algorithm simulates the natural musical improvisation process where the musicians aim to achieve a near-perfect state of harmony [127]. In the HS algorithm, the candidate solution population is known as harmony memory (HM), where every single solution in solution space is referred to as "harmony," which belongs to the "n"-dimensional vector. Though HS has been successfully used for numerous applications across diverse domains, its applications for signal control optimization are limited. In a recent study, Gao et al. applied to HS in addition to four others metaheuristics for traffic signal scheduling (TSS) problems [121]. Experiments were conducted on real-time data from signalized intersections in Singapore to examine the performance of proposed metaheuristics. The authors considered heterogeneous traffic conditions. Simulation results proved the adequacy of all algorithms; however, the hybrid algorithm (ABC-LS) outperformed other techniques in terms of solution quality.

In another study, Ceylan and Ceylan adopted a hybrid harmony search algorithm and TRANSYT hill-climbing algorithm (HSHC-TRANS) for solving stochastic equilibrium network design (SEQND) in the context of optimal traffic signal setting problems [128]. The effectiveness of HSHC-TRANS was evaluated against HS and GA in terms of network performance index (PI). Results showed that the proposed hybrid model yielded about 11% in the network's PI compared to the GA-based model. In another study, Gao et al. addressed the urban traffic signal scheduling problem (TSSP) using a discrete harmony search (DHS) with an ensemble of local search [115]. The primary objective was to minimize the network-wide total delay under a pre-defined finite horizon. Extensive simulation experiments were carried out using traffic data from a partial transport network in Singapore. Comparative analysis showed that the HS algorithm as a meta-heuristic achieved better performance compared to fixed-cycle traffic signal control (FCSC). Dellorco et al. also investigated the applicability of HS for signal control optimization on the two-junction network with the fixed flow on the links [116]. A comparative analysis of HS with GA and HC algorithms showed that HS resulted in a better network's PI compared to its counterparts. Afterward, the validity of the proposed HS algorithm was assessed by applying it to a test network.

6.2 Jaya algorithm

The Jaya algorithm is a recently proposed metaheuristic initially introduced by R.V. Rao [129]. The word Jaya comes from Sanskrit, which means "victory." In the Jaya algorithm, the search strategy always attempts to be victorious by reaching the optimal and best solution, and thus it is named "Jaya." It is arguably one of the simplest and easy-to-implement metaheuristics. The main benefit of Jaya for optimization problems lies in the fact that this algorithm requires only common controlling parameters such as population size and the number of iterations and does not require any additional algorithm-specific constraints/parameters. While this algorithm has been successfully used for several scheduling and optimization problems in recent years, its applications in the domain of traffic scheduling and management are relatively scarce.

A recent study conducted by Gao et al. compared the performance of Jaya algorithms with other metaheuristics (like water cycle algorithm (WCO), genetic algorithm (GA), artificial bee colony, and harmony search (HS), and hybrid ABC-LS) for solving traffic light scheduling problem [121]. Simulation results showed all the algorithms achieved competitive results; however, the hybrid algorithm attained better accuracy and convergence. The proposed models were also tested on real-time traffic and phase data from a network of intersections in the Jurong area of Singapore. In another study, the authors proposed an improved Jaya algorithm for solving traffic light optimization problems in the context of large-scale urban transport networks [122]. The chosen performance index was to minimize the network-wide total traffic delay within a given time horizon. To enhance the search performance in the local search space, a neighborhood search operator was proposed. Experiments were carried out using traffic data for a case study from the Singapore transport network. Study results demonstrated the robustness and better performance of proposed improved Jaya algorithms against standard Jaya algorithm and exiting traffic light control scheme. In another follow-up study, Gao et al. studied large-scale urban traffic lights scheduling problems using three different metaheuristics, namely Jaya, WCO, and HS [117]. The objective function was to optimize the delay time of all vehicles network-wise under a fixed time horizon. This study also proposed a feature search operator (FSO) to improve the search performance of proposed metaheuristics. To examine the efficacy of proposed methods, experiments were carried out using real-time traffic data. It was concluded that metaheuristic-based traffic control could significantly improve the network performance compared to existing traffic control strategies. Numerical simulation results showed that in comparison to feature-based search (FBS), operator for all algorithms improved the total vehicle delay time by more than 26% in their worst case scenarios.

Figure 7a depicts the relationships between total network delay time (sec) and sampling intervals for a typical urban traffic network with 100 junctions from the west Jurong area in Singapore [117]. Minimum (min.), average (avg.)

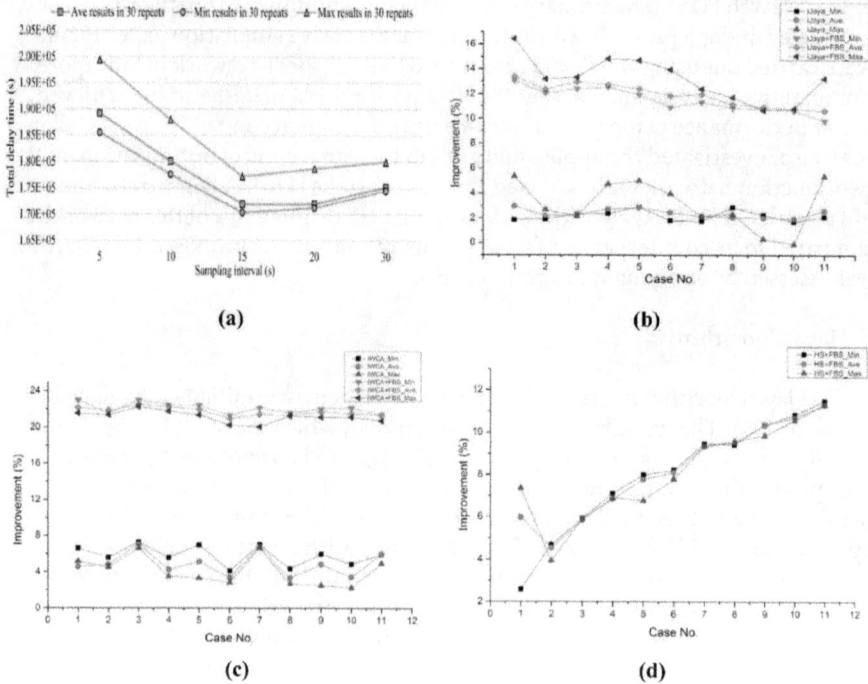

Figure 7.
(a) Results comparison with different sampling times for network of 100 junctions, (b) the % improvement of iJaya and iJaya+FBS with standard Jaya, (c) the % improvement IWCA and IWCA+FBS with standard Jaya, (d) the % improvement HS + FBS and standard HS. Ref. [117].

and maximum (max.) total delay values each for 30 repeats and five sampling intervals (5, 10, 15, 20, and 30 sec) are reported. It is evident from the results that a sampling period of 15 seconds yielded the best results, which were then adopted for subsequent experiments. **Figure 7b** shows the relative percentage improvement in network performance (reduction in network delay) for standard Jaya algorithm with improved Jaya (iJaya), and Jaya with FBS operator (iJaya+FBS) for a sample 11 cases of traffic network from the same study [117]. Compared to standard Jaya, the iJaya yielded the improvements in range for 0–6% for min., avg., and max. Results, while iJaya+FBS algorithm resulted in corresponding improvement values between 9 and 11%. **Figure 7c** depicts the percentage improvement of IWCA and IWCA+FBS algorithms relative to standard WCA optimizer. The IWCA improved the standard WCA in terms of min., avg., and max. Results for 11 test cases in the range of 2–8%, while the corresponding improvement for IWCA+FBS algorithm is approximately 20–24%. **Figure 7d** shows the network performance improvement of standard HS and HS + FBS algorithms for the same network of traffic junctions [117]. The improvement for HS + FBS algorithm compared to standard HS optimizer are between 2 and 12% for min., avg., and max. Results for the considered cases.

Figure 8 presents the graphical comparison among the three optimization algorithms (iJaya+FBS, IWCA+FBS, and HS + FBS) in terms of the average relative percentage deviation (ARPD) of the resulting network delay time values [117]. It is clear from the results that the IWCA+FBS algorithm with an average delay reduction of 28.54% outperformed the iJaya+FBS and HS + FBS having the corresponding values of 28.22% and 27.84%, respectively. Further, all the algorithms yielded an improvement of at least 26% in the worst-case scenarios.

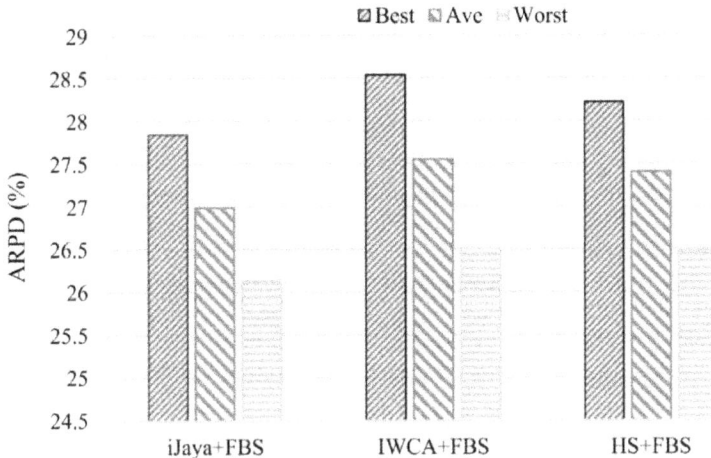

Figure 8.
ARPD improvements comparison for different optimizers. Reprinted with permission from Ref. [117] copyright (2021), Elsevier Ltd.

6.3 Water cycle algorithm (WCA)

The water cycle algorithm (WCA) is another recently proposed metaheuristic whose search mechanism is inspired by the natural water cycle process, where streams and rivers flow down the hill to reach the sea [130]. The surface run-off model is imitated in WCA for updating the current candidate solutions and the generation of new offspring. The effectiveness of WCA has been explored for various applications such as truss structures, constrained and unconstrained engineering design problems [130–133]. However, very few studies have used WCO for traffic control, management, and optimization.

A recent study by Gao et al. proposed the application WCO for traffic signal scheduling and optimization based on actual traffic data from a case study in Singapore [121]. WCO was compared with four other metaheuristics and a hybrid algorithm (ABC-LS), considering the network delay as the main optimization objective. Numerical simulation results proved the benefits of adopting metaheuristic-based traffic control strategies instead of existing fixed traffic light schemes. In another study, Gao et al. compared WCO with the Jaya algorithm and Harmony search using the field traffic data from the same transportation network. The performance metric minimized the network-wide total traffic delay within a given time horizon [117]. The study proposed a neighborhood search operator to enhance the search performance of all the algorithms in the local search space. Study results showed that WCA, with an average better improvement of in network-wide delay (28.54%), outperformed HS (28.22%) and Jaya algorithm (27.84%).

7. Conclusions, current challenges, and future research directions

Traffic control and management using metaheuristics have emerged as an effective solution to mitigate urban congestion. This study provided a comprehensive review of state-of-art research on traffic signal optimization using different metaheuristics approaches. The surveyed literature is categorized based on the nature of applied metaheuristics, i.e., swarm intelligence (SI) techniques, evolutionary algorithms, trajectory-based metaheuristics, and others. Although numerous metaheuristics have been employed for signal optimization, GA, PSO, ACO, and ABC algorithms have been widely explored. Various traffic signal parameters such as cycle length, green splits, offsets, and phasing sequence are considered decision variables to solve signal control optimization problems. Similarly, studies have considered several optimization objectives such as delay, number of stops, travel time, throughput, queue, fuel consumption, exhaust emissions to address the problem. Some studies have adopted single-objective optimization, while others have attempted to solve traffic signal control as a multi-objective optimization problem. However, little work has been done to understand the correlations between the conflicting objectives which is vital for traffic engineers and decision-makers to evaluate their relative importance. Based on the presented survey work, the following passages present key challenges, research gaps, and future research directions in this area.

- The review has shown that most of the previous works have adopted a single metaheuristic method for TSC optimization. However, very few studies have investigated the applicability of hybrid or ensemble metaheuristics for solving TSC optimization problems. In general, hybrid techniques are more useful than traditional metaheuristics. Hence, the application of hybrid metaheuristics for signal optimization could be a promising research direction.

- Traditional evolutionary algorithms and swarm intelligence optimizers could yield acceptable solutions. However, the performance of these optimization techniques may be compared with recent state-of-the-art optimization approaches such as Teaching Learning Based Optimization Algorithm (TLBOA), Gravitational Search Algorithms (GSA), Rock Hyraxes Swarm Optimization (RHSO), hyper-heuristics, which are not explored yet for traffic signal optimization problems.

- The literature review also noted that most previous studies were focused on single-objective optimization; however, traffic engineers often have to deal with multiple conflicting objectives to optimize the performance at the network level. Alternatively, for multiobjective optimization, the vast majority of existing works introduce weights for different objectives and consequently tackle signal optimization as a signal objective optimization problem. To optimize different performance indicators along optimal paretofront, multiple objectives have to be properly optimized. Developing an optimizer for multiobjective scenarios remains a challenging issue and is worth exploring in future studies.

- Objective functions based on energy consumption and exhaust emissions have become a topic of increasing interest in recent years. From the reviewed literature, it was concluded that various approximate fuel consumptions and emission models were used for signal control optimization. Application of such approximate models could lead to an un-realistic traffic light setting. Future studies should consider the calibration of fuel consumption and emission models for a given network.

- It was also evident from the presented literature that there is a shortage of research on statistical performance evaluation of proposed metaheuristics. Therefore, it would be interesting to explore the statistical analysis of such optimization strategies in terms of worst, average, and best results. Likewise, statistical significance tests may be conducted to compare the performance among various metaheuristics in solving signal optimization problems.

- Lack of appropriate validation protocol is another important issue. Some studies have employed mere traffic simulation platforms to assess the validity of applied metaheuristics, while others have used them for isolated intersection scenarios or small traffic networks. Network optimization has become popular in recent years. For achieving desired improvements at the network level, the methods should be tested for large-scale complex transportation networks.

- The surveyed literature also indicated that most previous studies considered only vehicular traffic and neglected the pedestrian traffic in solving the TSC problem using metaheuristics. It is important to consider all forms of traffic and driving systems to improve the overall efficiency of the transport system.

- The surveyed literature also revealed that many studies develop metaheuristic-based traffic control considering specific traffic demand conditions, neglecting the other potential scenarios. It is essential to consider all ranges of traffic flow conditions (undersaturated, saturated, and oversaturated flow conditions) and traffic disturbances in developing metaheuristic to address TSC optimization problems.

- The accuracy and reliability of the signal timing plan obtained using metaheuristics are highly dependent on the accuracy of traffic flow prediction models. In recent years, with rapid advances in computational power, big data technology has been successfully used for accurate traffic flow prediction. Therefore, the application of metaheuristics coupled with big data technology for traffic signal optimization appears to be another interesting research direction.

Acknowledgements

The authors acknowledge the support of the King Fahd University of Petroleum and Minerals, KFUPM, Dhahran Saudi Arabia, and Qassim University, Burudah, Saudi Arabia, for Supporting this study.

Conflict of interest

"The authors declare no conflict of interest."

Author details

Arshad Jamal[1*], Hassan M. Al-Ahmadi[1], Farhan Muhammad Butt[2],
Mudassir Iqbal[3,4], Meshal Almoshaogeh[5] and Sajid Ali[6]

1 Department of Civil and Environmental Engineering, King Fahd University of Petroleum and Minerals, Dhahran, Saudi Arabia

2 Department of Transportation and Traffic Engineering, College of Engineering, Imam Abdulrahman Bin Faisal University, Dammam, Saudi Arabia

3 Shanghai Key Laboratory for Digital Maintenance of Buildings and Infrastructure, School of Naval Architecture, Ocean and Civil Engineering, Shangai Jiao Tong University, Shanghai, China

4 Department of Civil Engineering, University of Engineering and Technology Peshawar, Pakistan

5 Department of Civil engineering, College of Engineering, Qassim University, Buraydah, Qassim, Saudi Arabia

6 Mechanical and Energy Engineering Department, Imam Abdulrahman Bin Faisal University, Dammam, KSA

*Address all correspondence to: arshad.jamal@kfupm.edu.sa

IntechOpen

References

[1] M. Zahid, Y. Chen, A. Jamal, and Q. M. Memon, *Short Term Traffic State Prediction via Hyperparameter Optimization Based Classifiers*, vol. 20. 2020. doi: 10.3390/s20030685.

[2] M. Zahid, Y. Chen, and A. Jamal, "Freeway Short-Term Travel Speed Prediction Based on Data Collection Time-Horizons : A Fast Forest Quantile Regression Approach," Sustainability, vol. 12, no. 646, pp. 1-19, 2020, doi: doi:10.3390/su12020646.

[3] M. Dotoli, M. P. Fanti, and C. Meloni, "A signal timing plan formulation for urban traffic control," Control engineering practice, vol. 14, no. 11, pp. 1297-1311, 2006.

[4] Cambridge Systematics, Ed., "Traffic Congestion and Reliability: Trends and Advanced Strategies for Congestion Mitigation," no. FHWA-HOP-05-064, Sep. 2005, [Online]. Available: https://rosap.ntl.bts.gov/view/dot/20656

[5] C. Shirke, N. Sabar, E. Chung, and A. Bhaskar, "Metaheuristic approach for designing robust traffic signal timings to effectively serve varying traffic demand," Journal of Intelligent Transportation Systems, pp. 1-17, 2021.

[6] C.-J. Lan, "New optimal cycle length formulation for pretimed signals at isolated intersections," Journal of transportation engineering, vol. 130, no. 5, pp. 637-647, 2004.

[7] L. D. Han and J.-M. Li, "Short or long—Which is better? Probabilistic approach to cycle length optimization," Transportation Research Record, vol. 2035, no. 1, pp. 150-157, 2007.

[8] P. B. Hunt, D. I. Robertson, R. D. Bretherton, and M. C. Royle, "The SCOOT on-line traffic signal optimisation technique," *Traffic Engineering & Control*, vol. 23, no. 4, 1982.

[9] A. G. Sims and K. W. Dobinson, "The Sydney coordinated adaptive traffic (SCAT) system philosophy and benefits," IEEE Transactions on vehicular technology, vol. 29, no. 2, pp. 130-137, 1980.

[10] L. John, M. D. Kelson, and N. H. Gartner, "A Versatile Program for Setting Signals on Arteries and Triangular Networks," Transp. Res. Rec. J. Transp. Res. Board, vol. 795, pp. 40-46, 1981.

[11] J.-J. Henry, J. L. Farges, and J. Tuffal, "The PRODYN real time traffic algorithm," in *Control in Transportation Systems*, Elsevier, 1984, pp. 305-310.

[12] D. I. Robertson, "'TANSYT'METHOD FOR AREA TRAFFIC CONTROL," *Traffic Engineering & Control*, vol. 8, no. 8, 1969.

[13] S. Sen and K. L. Head, "Controlled optimization of phases at an intersection," Transportation science, vol. 31, no. 1, pp. 5-17, 1997.

[14] N. H. Gartner, *OPAC: A demand-responsive strategy for traffic signal control*. 1983.

[15] S. Chiu and S. Chand, "Adaptive Traffic Signal Control Using Fuzzy Logic," in *Proceedings*. The First IEEE Regional Conference on Aerospace Control Systems, 1993, pp. 122-126. doi: 10.1109/AEROCS.1993.720907.

[16] R. Akcelik, *Traffic signals: capacity and timing analysis*. 1981.

[17] H. C. Manual, "HCM2010," *Transportation Research Board, National Research Council, Washington, DC*, p. 1207, 2010.

[18] S. Göttlich, A. Potschka, and U. Ziegler, "Partial outer convexification

for traffic light optimization in road networks," SIAM Journal on Scientific Computing, vol. 39, no. 1, pp. B53–B75, 2017.

[19] K. Aboudolas, M. Papageorgiou, and E. Kosmatopoulos, "Store-and-forward based methods for the signal control problem in large-scale congested urban road networks," Transportation Research Part C: Emerging Technologies, vol. 17, no. 2, pp. 163-174, 2009.

[20] S. Voß, "Meta-heuristics: The state of the art," in *Workshop on Local Search for Planning and Scheduling*, 2000, pp. 1-23.

[21] J. Lee, B. Abdulhai, A. Shalaby, and E.-H. Chung, "Real-time optimization for adaptive traffic signal control using genetic algorithms," Journal of Intelligent Transportation Systems, vol. 9, no. 3, pp. 111-122, 2005.

[22] M. M. Abbas and A. Sharma, "Multiobjective plan selection optimization for traffic responsive control," Journal of transportation engineering, vol. 132, no. 5, pp. 376-384, 2006.

[23] L. Wu, Y. Wang, X. Yuan, and Z. Chen, "Multiobjective optimization of HEV fuel economy and emissions using the self-adaptive differential evolution algorithm," IEEE Transactions on vehicular technology, vol. 60, no. 6, pp. 2458-2470, 2011.

[24] Z. Guangwei, G. Albert, and L. D. Sherr, "Optimization of adaptive transit signal priority using parallel genetic algorithm," Tsinghua Science and Technology, vol. 12, no. 2, pp. 131-140, 2007.

[25] E. Doğan and A. P. Akgüngör, "Optimizing a fuzzy logic traffic signal controller via the differential evolution algorithm under different traffic scenarios," Simulation, vol. 92, no. 11, pp. 1013-1023, 2016.

[26] Z. Cakici and Y. S. Murat, "A Differential Evolution Algorithm-Based Traffic Control Model for Signalized Intersections," *Advances in Civil Engineering*, vol. 2019, 2019.

[27] S. Zhou, X. Yan, and C. Wu, "Optimization Model for Traffic Signal Control with Environmental Objectives," in *2008 Fourth International Conference on Natural Computation*, Jinan, Shandong, China, 2008, pp. 530-534. doi: 10.1109/ICNC.2008.494.

[28] W. Kou, X. Chen, L. Yu, and H. Gong, "Multiobjective optimization model of intersection signal timing considering emissions based on field data: A case study of Beijing," Journal of the Air and Waste Management Association, vol. 68, no. 8, pp. 836-848, 2018, doi: 10.1080/10962247. 2018.1454355.

[29] A. Jamal, M. T. Rahman, H. M. Al-Ahmadi, I. M. Ullah, and M. Zahid, "Intelligent Intersection Control for Delay Optimization: Using Meta-Heuristic Search Algorithms," *Sustainability*, vol. 12, no. 5, p. 1896, 2020.

[30] Z. Zhou and M. Cai, "Intersection signal control multi-objective optimization based on genetic algorithm," Journal of Traffic and Transportation Engineering (English Edition), vol. 1, no. 2, pp. 153-158, Apr. 2014, doi: 10.1016/ S2095-7564(15)30100-8.

[31] Yunrui Bi, Dipti Srinivasan, Xiaobo Lu, Zhe Sun and W. Zeng, "Type-2 fuzzy multi-intersection traffic signal control with differential evolution optimization," *Expert Systems with Applications*, pp. 7338-7349., 2014, doi: DOI:https://doi.org/10.1016/j.

[32] J. Kwak, B. Park, and J. Lee, "Evaluating the impacts of urban corridor traffic signal optimization on vehicle emissions and fuel

consumption," Transportation Planning and Technology, vol. 35, no. 2, pp. 145-160, Mar. 2012, doi: 10.1080/03081060.2011.651877.

[33] L. Adacher and A. Gemma, "A robust algorithm to solve the signal setting problem considering different traffic assignment approaches," International Journal of Applied Mathematics and Computer Science, vol. 27, no. 4, pp. 815-826, 2017.

[34] Y. Li, L. Yu, S. Tao, and K. Chen, "Multi-Objective Optimization of Traffic Signal Timing for Oversaturated Intersection," Mathematical Problems in Engineering, vol. 2013, pp. 1-9, 2013, doi: 10.1155/2013/182643.

[35] M. Al-Turki, A. Jamal, H. M. Al-Ahmadi, M. A. Al-Sughaiyer, and M. Zahid, "On the Potential Impacts of Smart Traffic Control for Delay, Fuel Energy Consumption, and Emissions: An NSGA-II-Based Optimization Case Study from Dhahran, Saudi Arabia," *Sustainability*, vol. 12, no. 18, p. 7394, 2020.

[36] M. D. Foy and R. F. Benekohal, "Signal timing determination using genetic algorithms," Transportation Reserach Record, *1365*, pp. 108-115, 1993.

[37] Q. Liu and J. Xu, "Traffic signal timing optimization for isolated intersections based on differential evolution bacteria foraging algorithm," Procedia-Social and Behavioral Sciences, vol. 43, pp. 210-215, 2012.

[38] E. Ricalde and W. Banzhaf, "Evolving adaptive traffic signal controllers for a real scenario using genetic programming with an epigenetic mechanism," in *2017 16th IEEE International Conference on Machine Learning and Applications (ICMLA)*, 2017, pp. 897-902.

[39] J. Holland, "Adaptation in natural and artificial systems: an introductory

analysis with application to biology," *Control and artificial intelligence*, 1975.

[40] L. Jian, "Multi-objective optimisation of traffic signal control based on particle swarm optimisation," International Journal of Grid and Utility Computing, vol. 11, no. 4, pp. 547-553, 2020.

[41] H. Yang and D. Luo, "Acyclic Real-Time Traffic Signal Control Based on a Genetic Algorithm," Cybernetics and Information Technologies, vol. 13, no. 3, pp. 111-123, Sep. 2013, doi: 10.2478/cait-2013-0029.

[42] M. Liu, Y. Oeda, and T. Sumi, "Multi-Objective Optimization of Intersection Signal Time Based on Genetic Algorithm," Memoirs of the Faculty of Engineering, Kyushu University, vol. 78, no. 4, pp. 14-23, 2018.

[43] J. Guo, Y. Kong, Z. Li, W. Huang, J. Cao, and Y. Wei, "A model and genetic algorithm for area-wide intersection signal optimization under user equilibrium traffic," Mathematics and Computers in Simulation, vol. 155, pp. 92-104, 2019.

[44] H. Dezani, N. Marranghello, and F. Damiani, "Genetic algorithm-based traffic lights timing optimization and routes definition using Petri net model of urban traffic flow," IFAC Proceedings Volumes, vol. 47, no. 3, pp. 11326-11331, 2014.

[45] M. K. Tan, H. S. E. Chuo, R. K. Y. Chin, K. B. Yeo, and K. T. K. Teo, "Optimization of traffic network signal timing using decentralized genetic algorithm," in *2017 IEEE 2nd International Conference on Automatic Control and Intelligent Systems (I2CACIS)*, 2017, pp. 62-67.

[46] K. V. Price, "Differential evolution," in *Handbook of optimization*, Springer, 2013, pp. 187-214.

[47] E. Korkmaz and A. P. AKGÜNGÖR, "Delay estimation models for signalized intersections using differential evolution algorithm," *Journal of Engineering Research*, vol. 5, no. 3, 2017.

[48] H. Ceylan, "Optimal Design of Signal Controlled Road Networks Using Differential Evolution Optimization Algorithm," Mathematical Problems in Engineering, vol. 2013, p. 696374, 2013, doi: 10.1155/2013/696374.

[49] W.-L. Liu, Y.-J. Gong, W.-N. Chen, and J. Zhang, "EvoTSC: An evolutionary computation-based traffic signal controller for large-scale urban transportation networks," Applied Soft Computing, vol. 97, p. 106640, 2020.

[50] M. Zhang, S. Zhao, J. Lv, and Y. Qian, "Multi-phase urban traffic signal real-time control with multi-objective discrete differential evolution," in *2009 International Conference on Electronic Computer Technology*, 2009, pp. 296-300.

[51] W. Banzhaf, P. Nordin, R. E. Keller, and F. D. Francone, *Genetic programming: an introduction*, vol. 1. Morgan Kaufmann Publishers San Francisco, 1998.

[52] L. Vanneschi and R. Poli, "24 Genetic Programming–Introduction, Applications, Theory and Open Issues," 2012.

[53] D. J. Montana and S. Czerwinski, "Evolving control laws for a network of tra c signals," *Koza et al*, vol. 1492.

[54] E. Ricalde, "A genetic programming system with an epigenetic mechanism for traffic signal control," *arXiv preprint arXiv:1903.03854*, 2019.

[55] L. Bieker, D. Krajzewicz, A. Morra, C. Michelacci, and F. Cartolano, "Traffic simulation for all: a real world traffic scenario from the city of Bologna," in *Modeling Mobility with Open Data*, Springer, 2015, pp. 47-60.

[56] E. Ricalde and W. Banzhaf, "A genetic programming approach for the traffic signal control problem with epigenetic modifications," in *European Conference on Genetic Programming*, 2016, pp. 133-148.

[57] J. He and Z. Hou, "Ant colony algorithm for traffic signal timing optimization," Advances in Engineering Software, vol. 43, no. 1, pp. 14-18, Jan. 2012, doi: 10.1016/j. advengsoft.2011.09.002.

[58] B. Trabelsi, S. Elkosantini, and S. Darmoul, "Traffic Control at Intersections Using Artificial Immune System Approach," *9th International Conference of Modeling, Optimization and Simulation - MOSIM'12*, 2012.

[59] S. S. Sabry and N. M. Kaittan, "Grey wolf optimizer based fuzzy-PI active queue management design for network congestion avoidance," Indonesian Journal of Electrical Engineering and Computer Science, vol. 18, no. 1, pp. 199-208, 2020.

[60] H. Zhao, R. He, and J. Su, "Multi-objective optimization of traffic signal timing using non-dominated sorting artificial bee colony algorithm for unsaturated intersections," *Archives of Transport*, vol. 46, 2018.

[61] H. Min, "On Signal Timing Optimization in Isolated Intersection Based on the Improved Ant Colony Algorithm," in *International Symposium on Parallel Architecture, Algorithm and Programming*, 2017, pp. 439-443.

[62] K. Jintamuttha, B. Watanapa, and N. Charoenkitkarn, "Dynamic traffic light timing optimization model using bat algorithm," in *2016 2nd International Conference on Control Science and Systems Engineering (ICCSSE)*, 2016, pp. 181-185.

[63] S. Araghi, A. Khosravi, and D. Creighton, "Intelligent cuckoo search

optimized traffic signal controllers for multi-intersection network," Expert Systems with Applications, vol. 42, no. 9, pp. 4422-4431, 2015.

[64] M. A. Gökçe, E. Öner, and G. Işık, "Traffic signal optimization with Particle Swarm Optimization for signalized roundabouts," SIMULATION, vol. 91, no. 5, pp. 456-466, May 2015, doi: 10.1177/0037549715581473.

[65] C. Dong, S. Huang, and X. Liu, "Urban Area Traffic Signal Timing Optimization Based on Sa-PSO," in *2010 International Conference on Artificial Intelligence and Computational Intelligence*, Sanya, China, Oct. 2010, pp. 80-84. doi: 10.1109/AICI.2010.257.

[66] S. Srivastava and S. K. Sahana, "Application of bat algorithm for transport network design problem," *Applied Computational Intelligence and soft computing*, vol. 2019, 2019.

[67] A. C. Olivera, J. M. García-Nieto, and E. Alba, "Reducing vehicle emissions and fuel consumption in the city by using particle swarm optimization," Applied Intelligence, vol. 42, no. 3, pp. 389-405, 2015.

[68] W. Hao, C. Ma, B. Moghimi, Y. Fan, and Z. Gao, "Robust optimization of signal control parameters for unsaturated intersection based on tabu search-artificial bee colony algorithm," IEEE Access, vol. 6, pp. 32015-32022, 2018.

[69] V. C. SS, "A Multi-agent Ant Colony Optimization Algorithm for Effective Vehicular Traffic Management," in *International Conference on Swarm Intelligence*, 2020, pp. 640-647.

[70] S. Araghi, A. Khosravi, and D. Creighton, "Design of an optimal ANFIS traffic signal controller by using cuckoo search for an isolated intersection," in *2015 IEEE international conference on systems, man, and cybernetics*, 2015, pp. 2078-2083.

[71] N. Rida, M. Ouadoud, and A. Hasbi, "Ant colony optimization for real time traffic lights control on a single intersection," 2020.

[72] H. Jia, Y. Lin, Q. Luo, Y. Li, and H. Miao, "Multi-objective optimization of urban road intersection signal timing based on particle swarm optimization algorithm," *Advances in Mechanical Engineering*, vol. 11, no. 4, p. 168781401984249, Apr. 2019, doi: 10.1177/1687814019842498.

[73] H. S. E. Chuo, M. K. Tan, A. C. H. Chong, R. K. Y. Chin, and K. T. K. Teo, "Evolvable traffic signal control for intersection congestion alleviation with enhanced particle swarm optimisation," in *2017 IEEE 2nd International Conference on Automatic Control and Intelligent Systems (I2CACIS)*, 2017, pp. 92-97.

[74] Y. Qian, C. Wang, H. Wang, and Z. Wang, "The optimization design of urban traffic signal control based on three swarms cooperative-particle swarm optimization," in *2007 IEEE International Conference on Automation and Logistics*, 2007, pp. 512-515.

[75] A. Louati, S. Darmoul, S. Elkosantini, and L. ben Said, "An artificial immune network to control interrupted flow at a signalized intersection," Information Sciences, vol. 433, pp. 70-95, 2018.

[76] J. Kwiecień and B. Filipowicz, "Firefly algorithm in optimization of queueing systems," Bulletin of the Polish Academy of Sciences. Technical Sciences, vol. 60, no. 2, pp. 363-368, 2012.

[77] S. A. Çeltek, A. Durdu, and M. E. M. Alı, "Real-time traffic signal control with swarm optimization methods," Measurement, vol. 166, p. 108206, 2020.

[78] R. K. Abushehab, B. K. Abdalhaq, and B. Sartawi, "Genetic vs. particle swarm optimization techniques for traffic light signals timing," in *2014 6th International Conference on Computer Science and Information Technology (CSIT)*, 2014, pp. 27-35.

[79] N. Angraeni, M. A. Muslim, and A. T. Putra, "Traffic control optimization on isolated intersection using fuzzy neural network and modified particle swarm optimization," in *Journal of Physics: Conference Series*, 2019, vol. 1321, no. 3, p. 032023.

[80] J. Garcia-Nieto, A. C. Olivera, and E. Alba, "Optimal cycle program of traffic lights with particle swarm optimization," IEEE Transactions on Evolutionary Computation, vol. 17, no. 6, pp. 823-839, 2013.

[81] K.-R. Lo, Y. City, and T. County, "TRAFFIC SIGNAL CONTROL BASED ON PARTICLE SWARM OPTIMIZATION," p. 13.

[82] Y. Wei, Q. Shao, Y. Han, and B. Fan, "Intersection signal control approach based on pso and simulation," in *2008 Second International Conference on Genetic and Evolutionary Computing*, 2008, pp. 277-280.

[83] I. G. P. S. Wijaya, K. Uchimura, and G. Koutaki, "Traffic light signal parameters optimization using particle swarm optimization," in *2015 International Seminar on Intelligent Technology and Its Applications (ISITIA)*, 2015, pp. 11-16.

[84] M. Dorigo, M. Birattari, and T. Stutzle, "Ant colony optimization," IEEE computational intelligence magazine, vol. 1, no. 4, pp. 28-39, 2006.

[85] R. Putha, L. Quadrifoglio, and E. Zechman, "Comparing Ant Colony Optimization and Genetic Algorithm Approaches for Solving Traffic Signal Coordination under Oversaturation Conditions," Computer-Aided Civil and Infrastructure Engineering, vol. 27, no. 1, pp. 14-28, 2012, doi: 10.1111/j.1467-8667.2010.00715.x.

[86] H. Yu, R. Ma, and H. M. Zhang, "Optimal traffic signal control under dynamic user equilibrium and link constraints in a general network," Transportation research part B: methodological, vol. 110, pp. 302-325, 2018.

[87] S. Haldenbilen, O. Baskan, and C. Ozan, "An ant colony optimization algorithm for area traffic control," Ant colony optimization–techniques and applications, pp. 87-105, 2013.

[88] L. Li, Y. Ma, B. Wang, H. Dong, and Z. Zhang, "Research on traffic signal timing method based on ant colony algorithm and fuzzy control theory," Proceedings of Engineering and Technology Innovation, vol. 11, p. 21, 2019.

[89] M. R. Jabbarpour, H. Malakooti, R. M. Noor, N. B. Anuar, and N. Khamis, "Ant colony optimisation for vehicle traffic systems: applications and challenges," International Journal of Bio-Inspired Computation, vol. 6, no. 1, pp. 32-56, 2014.

[90] D. Renfrew and X.-H. Yu, "Traffic signal control with swarm intelligence," in *2009 Fifth International Conference on Natural Computation*, 2009, vol. 3, pp. 79-83.

[91] D. Renfrew and X.-H. Yu, "Traffic signal optimization using ant colony algorithm," in *The 2012 International Joint Conference on Neural Networks (IJCNN)*, 2012, pp. 1-7.

[92] S. Srivastava and S. K. Sahana, "Nested hybrid evolutionary model for traffic signal optimization," Applied Intelligence, vol. 46, no. 1, pp. 113-123, 2017.

[93] D. Karaboga, B. Akay, and C. Ozturk, "Artificial bee colony (ABC)

optimization algorithm for training feed-forward neural networks," in *International conference on modeling decisions for artificial intelligence*, 2007, pp. 318-329.

[94] D. Karaboga and B. Basturk, "On the performance of artificial bee colony (ABC) algorithm," Applied soft computing, vol. 8, no. 1, pp. 687-697, 2008.

[95] D. Karaboga and B. Basturk, "A powerful and efficient algorithm for numerical function optimization: artificial bee colony (ABC) algorithm," Journal of global optimization, vol. 39, no. 3, pp. 459-471, 2007.

[96] M. Dell'Orco, Ö. Başkan, and M. Marinelli, "Artificial bee colony-based algorithm for optimising traffic signal timings," in *Soft Computing in Industrial Applications*, Springer, 2014, pp. 327-337.

[97] X.-S. Yang and S. Deb, "Cuckoo search via Lévy flights," in *2009 World congress on nature & biologically inspired computing (NaBIC)*, 2009, pp. 210-214.

[98] X.-S. Yang and S. Deb, "Cuckoo search: recent advances and applications," Neural Computing and Applications, vol. 24, no. 1, pp. 169-174, 2014.

[99] X.-S. Yang, "A new metaheuristic bat-inspired algorithm," in *Nature inspired cooperative strategies for optimization (NICSO 2010)*, Springer, 2010, pp. 65-74.

[100] A. H. Gandomi, X.-S. Yang, A. H. Alavi, and S. Talatahari, "Bat algorithm for constrained optimization tasks," Neural Computing and Applications, vol. 22, no. 6, pp. 1239-1255, 2013.

[101] L. N. De Castro and J. Timmis, "An artificial immune network for multimodal function optimization," in *Proceedings of the 2002 Congress on Evolutionary Computation. CEC'02 (Cat. No. 02TH8600)*, 2002, vol. 1, pp. 699-704.

[102] S. Darmoul, S. Elkosantini, A. Louati, and L. B. Said, "Multi-agent immune networks to control interrupted flow at signalized intersections," Transportation Research Part C: Emerging Technologies, vol. 82, pp. 290-313, 2017.

[103] D. Moalla, S. Elkosantini, and S. Darmoul, "An artificial immune network to control traffic at a single intersectio," in *Proceedings of 2013 International Conference on Industrial Engineering and Systems Management (IESM)*, 2013, pp. 1-7.

[104] P. Negi, "Artificial immune system based urban traffic control," PhD Thesis, Texas A&M University, 2007.

[105] R. Galvan-Correa *et al.*, "Micro Artificial Immune System for Traffic Light Control," *Applied Sciences*, vol. 10, no. 21, p. 7933, 2020.

[106] Z. Qiao, L. Ke, G. Zhang, and X. Wang, "Adaptive collaborative optimization of traffic network signal timing based on immune-fireworks algorithm and hierarchical strategy," Applied Intelligence, pp. 1-17, 2021.

[107] X.-S. Yang, *Nature-inspired metaheuristic algorithms*. Luniver press, 2010.

[108] S. Goudarzi, M. N. Kama, M. H. Anisi, S. A. Soleymani, and F. Doctor, "Self-organizing traffic flow prediction with an optimized deep belief network for internet of vehicles," *Sensors*, vol. 18, no. 10, p. 3459, 2018.

[109] S. Mirjalili, S. M. Mirjalili, and A. Lewis, "Grey wolf optimizer," Advances in engineering software, vol. 69, pp. 46-61, 2014.

[110] X. Zhang, Q. Lin, W. Mao, S. Liu, Z. Dou, and G. Liu, "Hybrid Particle

Swarm and Grey Wolf Optimizer and its application to clustering optimization," Applied Soft Computing, vol. 101, p. 107061, 2021.

[111] T.-C. Teng, M.-C. Chiang, and C.-S. Yang, "A hybrid algorithm based on GWO and GOA for cycle traffic light timing optimization," in *2019 IEEE International Conference on Systems, Man and Cybernetics (SMC)*, 2019, pp. 774-779.

[112] Z. Li and P. Schonfeld, "Hybrid simulated annealing and genetic algorithm for optimizing arterial signal timings under oversaturated traffic conditions," Journal of advanced transportation, vol. 49, no. 1, pp. 153-170, 2015.

[113] Z.-R. Song, L.-L. Zang, and W.-X. Zhu, "Study on minimum emission control strategy on arterial road based on improved simulated annealing genetic algorithm," Physica A: Statistical Mechanics and its Applications, vol. 537, p. 122691, 2020.

[114] T. Oda, T. Otokita, T. Tsugui, and Y. Mashiyama, "Application of simulated annealing to optimization of traffic signal timings," IFAC Proceedings Volumes, vol. 30, no. 8, pp. 733-736, 1997.

[115] K. Gao, Y. Zhang, A. Sadollah, and R. Su, "Optimizing urban traffic light scheduling problem using harmony search with ensemble of local search," Applied Soft Computing, vol. 48, pp. 359-372, 2016.

[116] M. Dell'Orco, O. Baskan, and M. Marinelli, "A Harmony Search Algorithm approach for optimizing traffic signal timings," *PROMET-Traffic&Transportation*, vol. 25, no. 4, pp. 349-358, 2013.

[117] K. Gao, Y. Zhang, A. Sadollah, A. Lentzakis, and R. Su, "Jaya, harmony search and water cycle algorithms for

solving large-scale real-life urban traffic light scheduling problem," Swarm and evolutionary computation, vol. 37, pp. 58-72, 2017.

[118] T.-Y. Hu and L.-W. Chen, "Traffic signal optimization with greedy randomized tabu search algorithm," Journal of transportation engineering, vol. 138, no. 8, pp. 1040-1050, 2012.

[119] A. Karoonsoontawong and S. T. Waller, "Application of reactive tabu search for combined dynamic user equilibrium and traffic signal optimization problem," Transportation research record, vol. 2090, no. 1, pp. 29-41, 2009.

[120] M. A. Chentoufi and R. Ellaia, "A Hybrid Particle Swarm Optimization and Tabu Search algorithm for adaptive traffic signal timing optimization," in *2018 IEEE International Conference on Technology Management, Operations and Decisions (ICTMOD)*, 2018, pp. 25-30.

[121] K. Gao, Y. Zhang, R. Su, F. Yang, P. N. Suganthan, and M. Zhou, "Solving traffic signal scheduling problems in heterogeneous traffic network by using meta-heuristics," IEEE Transactions on Intelligent Transportation Systems, vol. 20, no. 9, pp. 3272-3282, 2018.

[122] K. Gao, Y. Zhang, A. Sadollah, and R. Su, "Jaya algorithm for solving urban traffic signal control problem," in *2016 14th International Conference on Control, Automation, Robotics and Vision (ICARCV)*, 2016, pp. 1-6.

[123] P. R. Lowrie and L. PR, "The Sydney co-ordinated adaptive traffic system: Principles, methodology, algorithms," 1982.

[124] F. Glover, "Tabu search and adaptive memory programming—advances, applications and challenges," in *Interfaces in computer science and operations research*, Springer, 1997, pp. 1-75.

[125] S. Kirkpatrick, C. D. Gelatt, and M. P. Vecchi, "Optimization by simulated annealing," science, vol. 220, no. 4598, pp. 671-680, 1983.

[126] R. A. Rutenbar, "Simulated annealing algorithms: An overview," IEEE Circuits and Devices magazine, vol. 5, no. 1, pp. 19-26, 1989.

[127] Z. W. Geem, J. H. Kim, and G. V. Loganathan, "A new heuristic optimization algorithm: harmony search," simulation, vol. 76, no. 2, pp. 60-68, 2001.

[128] H. Ceylan and H. Ceylan, "A hybrid harmony search and TRANSYT hill climbing algorithm for signalized stochastic equilibrium transportation networks," Transportation Research Part C: Emerging Technologies, vol. 25, pp. 152-167, 2012.

[129] R. Rao, "Jaya: A simple and new optimization algorithm for solving constrained and unconstrained optimization problems," International Journal of Industrial Engineering Computations, vol. 7, no. 1, pp. 19-34, 2016.

[130] H. Eskandar, A. Sadollah, A. Bahreininejad, and M. Hamdi, "Water cycle algorithm–A novel metaheuristic optimization method for solving constrained engineering optimization problems," Computers & Structures, vol. 110, pp. 151-166, 2012.

[131] A. Sadollah, H. Eskandar, A. Bahreininejad, and J. H. Kim, "Water cycle, mine blast and improved mine blast algorithms for discrete sizing optimization of truss structures," Computers & Structures, vol. 149, pp. 1-16, 2015.

[132] A. Sadollah, H. Eskandar, A. Bahreininejad, and J. H. Kim, "Water cycle algorithm with evaporation rate for solving constrained and unconstrained optimization problems,"

Applied Soft Computing, vol. 30, pp. 58-71, 2015.

[133] Y. Zhang *et al.*, "Application of an enhanced BP neural network model with water cycle algorithm on landslide prediction," Stochastic Environmental Research and Risk Assessment, pp. 1-19, 2020.

Chapter 6

A Comparison Study of PAPR Reduction in OFDM Systems Based on Swarm Intelligence Algorithms

Lahcen Amhaimar, Ali Elyaakoubi, Mohamed Bayjja,
Kamal Attari and Saida Ahyoud

Abstract

Optimization algorithms have been one of the most important research topics in Computational Intelligence Community. They are widely utilized mathematical functions that solve optimization problems in a variety of purposes via the maximization or minimization of a function. The swarm intelligence (SI) optimization algorithms are an active branch of Evolutionary Computation, they are increasingly becoming one of the hottest and most important paradigms, several algorithms were proposed for tackling optimization problems. The most respected and popular SI algorithms are Ant colony optimization (ACO) and particle swarm optimization (PSO). Fireworks Algorithm (FWA) is a novel swarm intelligence algorithm, which seems effective at finding a good enough solution of a complex optimization problem. In this chapter we proposed a comparison study to reduce the high PAPR (Peak-to-Average Power Ratio) in OFDM systems based on the swarm intelligence algorithms like simulated annealing (SA), particle swarm optimization (PSO), fireworks algorithm (FWA), and genetic algorithm (GA). It turns out from the results that some algorithms find a good enough solutions and clearly outperform the others candidates in both convergence speed and global solution accuracy.

Keywords: OFDM, PAPR, PTS, Swarm Intelligence, Fireworks Algorithm, GA, PSO

1. Introduction

In the last decade, Swarm Intelligence (SI) optimization algorithms attracted a great deal of attention and become popular among researchers from different fields and diverse domains working on optimization problems all over the world [1, 2]. The SI methods are increasingly becoming one of the most important research topics of evolutionary computation (EC).

In the past several years, fruitful achievements have been made in the Computational Intelligence researches areas, such as evolutionary computation [3–5], simulated annealing [6], artificial neural networks [7–9], tabu search [10], chaos computation [11], fuzzy logic and systems [12]. All these methods inspired by natural Behavior.

In general, swarm intelligence algorithms (SI) can be divided into two main categories, bio-inspired and non-bio-inspired. The first one includes particle swarm optimization (PSO) [13], ant colony optimization (ACO) [14], artificial bee algorithm (ABC) [15, 16], fish schooling search (FSS) [17], bacterial foraging optimization (BFO) [18], firefly algorithm-II [19], bat algorithm [20] and so forth. The second categories of non-bio-inspired algorithms includes fireworks algorithm (FWA) [21], brain storm optimization (BSO) [22], magnetic optimization algorithms [23] and water drops algorithm [24]. Each algorithm has some advantages in solving many optimization problems but among all these algorithms, PSO, FWA and GA [25, 26] are the most popular algorithms for searching optimal locations in a D-dimensional space.

This chapter aims to present a comparison study to resolve an optimization problem in orthogonal frequency division multiplexing (OFDM) system based on the important evolutionary algorithm in the literature.

The multicarrier modulation techniques like OFDM [27–29] provides a viable alternative to enhance the quality of service for data transmission over single carrier systems, it has various advantages and now being used in a number of wireless communication systems. However, the OFDM system still suffers from the high envelope fluctuations of the transmitted signal called the peak-to-average power ratio (PAPR). This main concern improves the complexity of nonlinear elements, decreases the efficiency of high power amplifiers (HPA), and causes out-of-band radiation with degradation of bit error rate (BER).

Partial transmit sequences (PTS) [30–32] is one of the most attractive technique and a promising scheme due to its efficiency in PAPR reduction, but it requires an exhaustive search to find the optimum phase factors, which causes high computational complexity increased with the number of phase and subblocks. In this paper we will try to present many novel algorithms and their efficient improvements combined with PTS scheme to reduce the PAPR and the computational complexity.

The rest of the chapter is organized as follows. In Section 2, OFDM system model and the PAPR problem is formulated, and then the principles of PTS techniques are introduced. In Section 3, we have introduced the paradigm of Swarm Intelligence algorithms, and outlined the technical details of some popular SI algorithms like FWA, PSO, and GA. We also discuss the characteristics, the framework of the FWA based PTS and some simulation results under this Section. while Sections 4 and 5 are devoted to the comparison study of computational complexity and conclusions successively.

2. Multicarrier modulation (OFDM) and PTS approach

2.1 PAPR in OFDM signal

Wireless communications systems have experienced explosive growth with the demand for high data rate, theses digital systems require each channel to operate at a specific frequency and with a specific bandwidth. OFDM systems are currently being implemented in some of the newest and most advanced communications systems due to its effectiveness in using the frequency spectrum. OFDM is a subset of frequency division multiplexing in which a single channel utilizes multiple sub-carriers on adjacent frequencies, this typical technique divides the effective spectrum channel to a number of orthogonal sub-channels and with equal bandwidth, each sub-channel handles independently with it's own data using individual subcarrier and the OFDM signal is the sum of all independent subcarriers. As a result, OFDM systems are able to maximize spectral efficiency without causing adjacent channel interference. OFDM signal generated by mapped the input data of binary sequences into complex data symbols called constellation, by a modulator (PSK, QPSK, QAM, etc.). Then, after serial to parallel conversion, N mapped symbols $X = [X_0, X_1,...,X_{N-1}]^T$ are fed to IDFT block to formed the time domain OFDM signal $x = [x_0,x_1,...,x_{N-1}]^T$. In the discrete time domain and with oversampled factor L = 4 The mathematical expression of the complex envelope of OFDM signal can be written as

$$x[n] = \frac{1}{\sqrt{N}} \sum_{k=0}^{N-1} X_k e^{\frac{j2\pi nk}{LN}}, 0 \leq n \leq LN - 1 \tag{1}$$

where N is the number of subcarriers and X_k is the nth complex symbol carried and transmitted by the kth subcarrier.

In the time domain, the transmit signals in an OFDM system can have high peak values since many subcarrier components are added via an inverse fast Fourier transformation (IFFT) operation. Compared to single-carrier systems, OFDM systems are known to have a high peak-to-average power ratio (PAPR), which decreases the signal-to-quantization noise ratio (SQNR) of the digital-analog convertor (DAC) and analog-digital convertor (ADC) while degrading the efficiency of the high power amplifier (HPA) in the transmitter.

The PAPR of a signal in discrete time is defined as the ratio between the maximum power and the average power of the complex OFDM signal, it can be expressed by the following formula [29]:

$$PAPR\{x[n]\} = \frac{\max\left\{|x[n]|^2\right\}}{E\left\{|x[n]|^2\right\}}, 0 \leq n \leq LN - 1 \tag{2}$$

where x[n] is given by (Eq. (1)) and E {.} denotes the expected value (Average power).

2.2 Partial transmit sequence (PTS)

When The PAPR reduction technique of Partial Transmit Sequence (PTS), was proposed in the framework of the continuity of the "Selective Mapping" technique [33]. It is based on the same principle as the SLM, with a multiple representation of the signal. The basic idea of this method has been described and detailed by S.H Muller and J.B Huber in [31, 34]. It consists in partitioning an input data block of N subcarriers into V subblocks of the same size with N/V subcarriers per each subblock. Each subcarrier allocated to the data transmitted in one sub-block will be set to zero in all others.

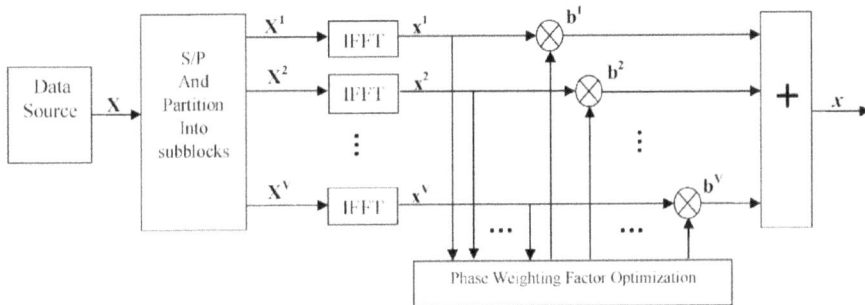

Figure 1.
Block diagram of PTS technique.

Once the V sub-blocks have been formed, the PTS technique applies a phase rotation optimization on each v sub block after IDFT to form the final signal at the lowest PAPR **Figure 1**.

The principle of the PTS technique is illustrated in **Figure 1**, where the algorithm is described as follows:

i. Firstly, after digital modulation, the symbols are subdivided into V sub-blocks, of equal size, such that the original signal is $X = \sum_{v=1}^{V} X^v$

ii. A phase shift is applied to all data symbols in each independent disjoint sub-blockX^v and the new frequency OFDM symbol is written as:

$$X = \sum_{v=1}^{V} X^v . b^v, b^v = e^{j\varphi^v}, v = 1, 2, ..., V. \tag{3}$$

iii. Subsequently, IFFT is applied on each sub-block to determine the modified OFDM symbol in the time domain.

$$x = IFFT\left\{ \sum_{v=1}^{V} b^v X^v \right\} = \sum_{v=1}^{V} b^v . IFFT\{X^v\} = \sum_{v=1}^{V} b^v x^v, \tag{4}$$

where the phase vectorb^v is chosen so that the PAPR can be minimized. It is optimized as follows:

$$[b^1, ... b^V] = \arg \min_{[b^1, ... b^V]} \left(\max_{n=0,1,...,N-1} \left| \sum_{v=1}^{V} b^v x^v [n] \right| \right) \tag{5}$$

In the practical application of wireless communication systems using the PTS approach, several drawbacks influence the performance of the PTS technique and

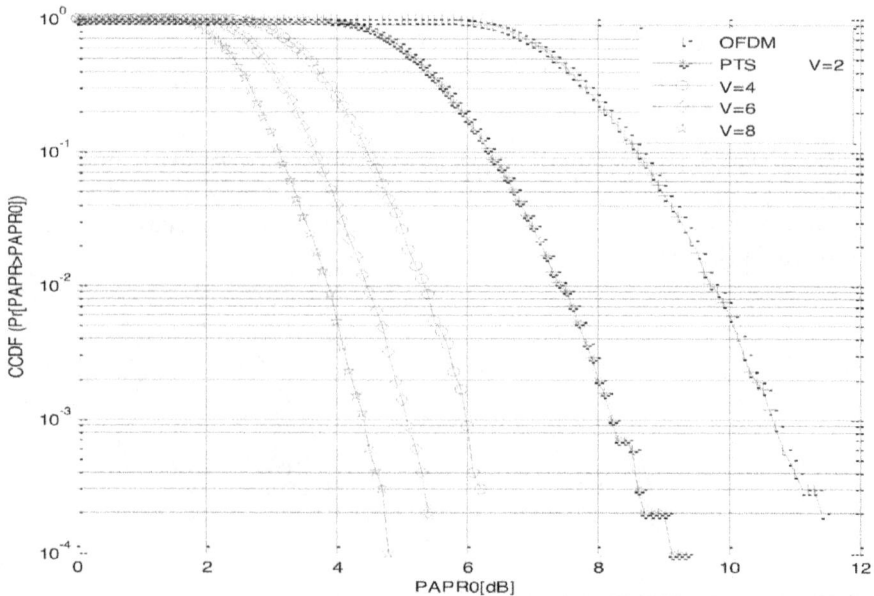

Figure 2.
PAPR reduction performance of PTS-OFDM when the number of sub-blocks varies.

increase its complexity. The PAPR performance will be improved as the number of subblocks V is increased (**Figure 2**). However, the complexity of the system also increased, to match the optimal phase weighting sequence for each input data sequence, W^V possible combinations should be checked (W number of phase factors). Moreover, the PTS technique requires the transmission of "Side Information" (SI) so that the receiver can identify the sequence that generated the lowest PAPR.

Figure 2 is an example of the PAPR for an OFDM signal with 64 subcarriers (802.11a), using a QPSK modulation and OFDM with PTS technique. From the above figure, the PTS method improves the PAPR performance as the number of sub-blocks increases. Several works aiming at complexity reduction, and several optimization algorithms have also been proposed to minimize the computational complexity, such as the Genetic Algorithm (GA) [25, 35], Particle Swarm Optimization (PSO) [36], simulating annealing (SA) [37] and so forth. We will compare theses algorithms with others new optimization methods in the next sections.

3. Swarm intelligence algorithms

Swarm intelligence algorithms have been widely used in many domains and attracted the attention of researchers working in optimization problems. It is one of the most important research topics in Computational Intelligence Community. The most of swarm intelligence algorithms have been inspired by some intelligent behaviors existing in nature like the collective behavior of a group of social insects (like bees, termites and wasps).

The most respected and popular SI algorithms are particle swarm optimization (PSO), which is inspired by the social behavior of bird flocking or fish schooling, fireworks algorithm (FWA) inspired by the fireworks explosion in the night sky, and ant colony optimization (ACO) which simulates the foraging behavior of ant colony. Nowadays, research efforts on SI are mainly devoted to algorithm design, problem solving, and applications, Hybrid algorithms and variants are actively proposed. The ACO, PSO, and the genetic algorithm (GA) are the most representative swarm intelligence algorithms applied to solve combinatorial optimization problems or used in real-parameter optimization.

3.1 Genetic algorithm (GA)

The genetic algorithm [25, 38], is an optimization algorithm based on techniques derived from genetics and natural evolution: crossovers, mutations, selection, etc. This optimization method has many advantages such as a good convergence, small computing time and high robustness. It can be used to select the optimal phase vector to reduce the PAPR [35], GA decreases the computational load of the PTS technique by searching a small piece of a set of possibilities instead of the whole set as in the classical technique. It searches for the extremum(s) of a function (PAPR function) defined on a space of dimension D, for example $[0\ 2\pi]$. To use it, we must have some basic elements.

The natural evolution is processed through three main steps as shown in **Figure 3**. First, a population with n chromosomes is generated randomly. Second, this population is exposed to some evolution mechanisms like crossover and mutation to form a new population with the hope of being better. Finally, some parts of the population are selected according to their fitness values (PAPR function) as in natural selection [38, 39]. The algorithm can be stopped when the maximum number of generations (max(Gen)) is reached, or meets a convergence requirement (a targeted PAPR). For more details, the reader can refer to [25].

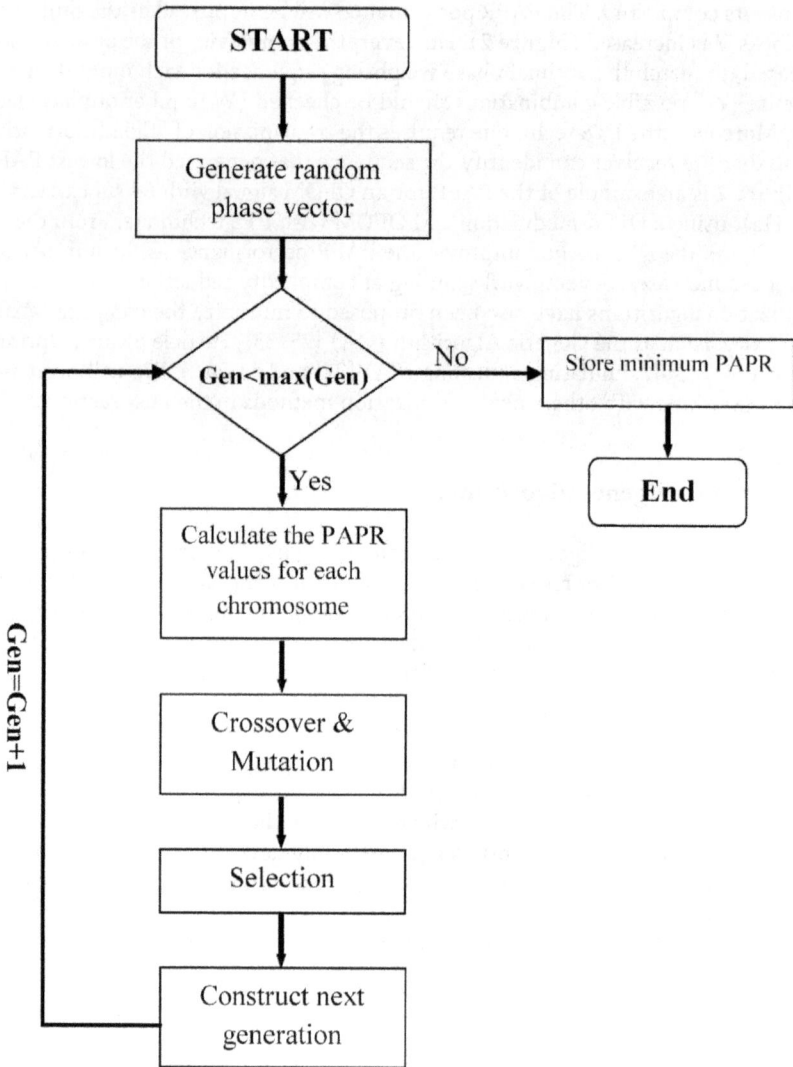

Figure 3.
Flow chart of genetic algorithm.

Parameter	Value
Number of generations (G)	5
Population (P)	5
Crossover rate (CR)	1.0
Mutation rate (MR)	0.05

Table 1.
GA simulation parameters.

The PTS method is combined with a GA to decrease the computational complexity. The basic configuration parameters of the genetic algorithm for the simulations are summarized in **Table 1**. The system uses $N = 64$ subcarriers, with QPSK
modulation. The signal is oversampled with the factor $L = 4$ and the weighting

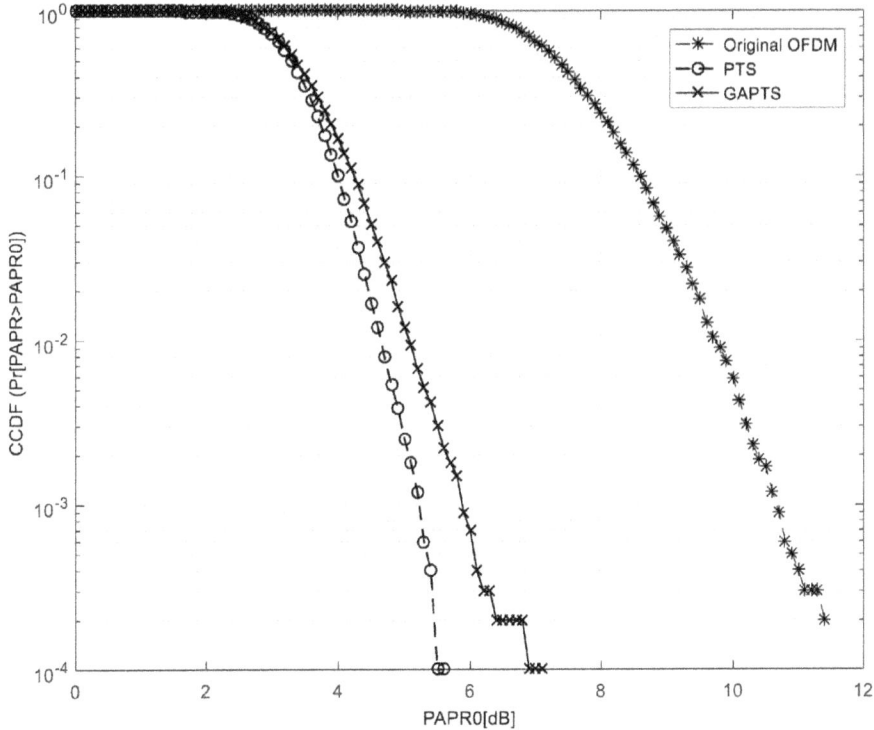

Figure 4.
Comparison of the PAPRo (dB) versus CCDF in OFDM systems for original PTS, and GA-PTS.

system uses a set of phase factors W = {1, −1, j, −j} to facilitate signal recovery. For all the results presented in this chapter, 10^4 OFDM symbols are generated, and the simulations are based on the IEEE 802.11/a standard.

Figure 4 shows the CCDF curves of OFDM system without PAPR reduction, the original PTS technique, and the GA-PTS technique. Although the PTS method has better PAPR than the GA-Proposed technique, the computational load of the PTS method is larger than GA-PTS.

3.2 Particle swarm optimization (PSO) based PTS

Particle swarm optimization (PSO) is a population-based global optimization technique put forward originally by Kennedy and E berhart in 1995 [36, 40–42], it is based on the research of bird and fish flock movement behavior. The PSO algorithm is a computational method used to solve non-linear continues problems and optimize many practical real life applications such as control reactive power, and Photovoltaic solar systems [43, 44], it has attracted the attention of researchers and a number of versions of PSO have been continuously proposed [45, 46].

In the basic particle swarm optimization algorithm, the population is called swarm and the individuals are called particles, so the PSO works by having a swarm of particles moved in the search space (D-dimensional) according to simple formulae until the optimal solution of the phase problem will be reached. During the movement of the population, each particle is characterized by two parameters: position and velocity. We used the PSO as an optimizer to reduce the PAPR by solving the phase factor problem in (Eq. (5)), the PSO algorithm evaluates each particle with the objective function of PAPR in (Eq. (2)).

During the optimization process; each solution is represented as a particle with a position vector x, referred to as a moving velocity and a phase weighting factor represented as v and b, respectively.

Thus for a K-dimensional optimization, the velocity and position of the i_{th} particle can be represented as $V_i = (v_{i,1}, v_{i,2}, . . ., v_{i,K})$ and $b_i = (b_{i,1}, b_{i,2}, ..., b_{i,K})$ respectively. Basically, each particle has its own best position referred to as pbest, $b_i^{Pb} = (b_{i,1}, b_{i,2}, ..., b_{i,K})$ corresponding to the individual best objective value obtained so far at time t, and the global best (gbest) particle is denoted by $b^{Gb} = (b_{g,1}, b_{g,2}, ..., b_{g,K})$, which represents the best particle found so far at time t in the entire swarm. So the expression of the new velocity $v_i(t + 1)$ for particle i is updated by

$$v_i(t + 1) = bv_i(t) + c_1 r_1 \left(b_i^{Pb}(t) - b_i(t) \right) + c_2 r_2 \left(b^G(t) - b_i(t) \right) \qquad (6)$$

where $v_i(t)$ is the old velocity of the particle i at time t, c_1, c_2 stand for acceleration constants and r_1, r_2 are random numbers between 0 and 1.

Based on the updated velocities (Eq. (6)), new position for particle i is computed according the following equation: $b_i(t + 1) = b_i(t) + v_i(t + 1)$.

In **Figure 5**, some results of the CCDF of the PAPR are simulated for the OFDM system with the same parameters used, in which the phase weight factor b = {±1, ±j} is used for PTS and GA-PTS, while the others algorithms like Standard PSO and simulated annealing used a search space [0 2π].

As we can see that the CCDF of the PAPR is well improved and the PSO-based PTS technique is capable of attaining a good PAPR performance, it is gradually promoted upon increasing and changing the research space dimension.

3.3 Simulated annealing (SA) based PTS

Simulated Annealing (SA) is an effective and general form of optimization. Over a number of years, the SA algorithm and its many extensions have been extensively

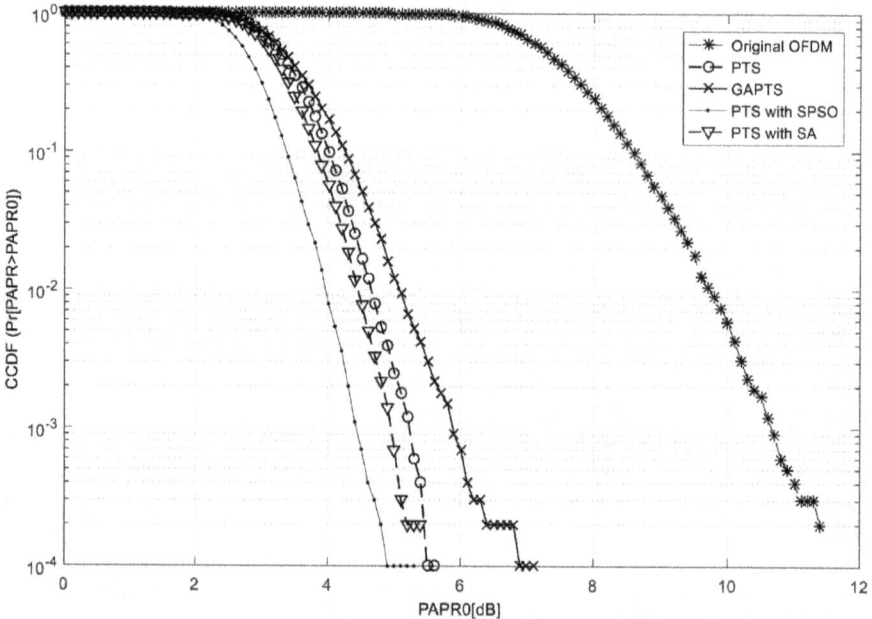

Figure 5.
CCDF of the PAPR with the PTS technique searched by SPSO, SA, and GA technique.

employed to solve a wide range of application domains, especially in combinatorial optimization problems [6, 47–49]. It is useful in finding global optima in the presence of large numbers of local optima. This characteristic makes the algorithm generic in the sense that it can be used to solve a variety of optimization problems without the need to change the basic structure of the computations. Over the last few years a number of variations to the original algorithm have been proposed, including parallel versions to speed up the rate of computations [50, 51].

"Annealing" refers to an analogy with thermodynamics, specifically with the way that metals cool and anneal. Simulated annealing uses the objective function of an optimization problem (PAPR in our case) instead of the energy of a material. The Simulated Annealing algorithm is a stochastic optimization method modeled on the behavior of condensed matter at low temperatures.

The Implementation of SA is surprisingly simple, at the outset, the system starts with a high T value, then annealing scheme is applied by slowly decreasing T according to some given procedure. The algorithm is basically hill-climbing except instead of picking the best move, it picks a random move at each T. If the selected move improves the solution (cost function of PAPR), then it is always accepted. Otherwise, in order to accept the states that do not improve the cost function (PAPR function), the algorithm makes the move anyway with some probability less than 1 depending on the PAPR reduction and T. This process randomizes the iterative improvement phase and avoid problems caused by moves that do not improve the solution in an attempt to reduce the probability of falling into a local minimum.

In our study we used SA based PTS algorithm to improve the search of phase factors for PAPR reduction in OFDM signals. QPSK modulation is employed with N = 64 subcarriers. The phase weighting factors W = $[0, 2\pi)$ have been used as in SPSO and 10^4 random OFDM symbols have been generated. In **Figure 5** Numerous computer simulations have been conducted to determine that the SA-PTS algorithm can improve PAPR performance better than GA and with a small difference with SPSO. (4,421 dB for SPSO and 4,948 dB for SA at CCDF = 10^{-3}).

3.4 Fireworks algorithm (FWA)

Fireworks algorithm (FWA) is an iterative swarm intelligence algorithm inspired by fireworks explosions in the sky at night, it was proposed by Y. Tan and Y. Zhu in 2010 [21] to searches for optimal solution of some optimization problems. FWA has attracted the attention of researchers and a number of versions of FWA have been continuously proposed [52–55].

The FWA is designed and implemented by simulating the explosion process of fireworks. It's made up of four key components, firstly explosive operator where two explosion processes are employed, explosion strength and explosion amplitude, secondly mutation operation, where the Gaussian mutation is the most widely used, thirdly mapping rule, and the most popular mapping rules are Mirror mapping rule and stochastic mapping rule, lastly as for selection strategy, there are distance-based selection and stochastic selection for keeping diversity of sparks.

3.4.1 Fireworks algorithm based PTS (FWA-PTS)

This section presents the basic principle, implementation and performance of FWA, aiming to develop this algorithm in a systematic and comprehensive way, and easily integrate it into an OFDM system to minimize PAPR. The approach is based on combining the PTS with the FWA to find the optimal phase vectors to reduce the PAPR with the least complexity [56, 57]. The objective function in this case is to minimize the PAPR of transmitted OFDM signals as follows:

$$\begin{cases} \text{Minimize } f_{Obj}(\mathbf{b}, x) = \sum_{v=1}^{V} b_v.x_v \\ \text{Subject to } b_v = \{e^{j\varphi_v}\}, v = 1, 2, \dots, V \end{cases} \tag{7}$$

where b_v represent the complex phase factors and the bounds of the potential space is defined by $0 \leq \varphi_v \leq 2\pi$.

Fireworks algorithm starts to run iteratively till the given termination conditions are met. When the FWA algorithm is initiated, a set of Sparks will fill the local space around the Firework, for a good optimization usually we started by five fireworks, and when we search for a point b_v satisfying $f_{obj}(b_v) = y$, we can continuously trigger Fireworks in the search space until a Spark checks the target or is close enough to the point b_v. **Figure 6** is depicting a rough framework of the search optimization algorithm of fireworks to find the best phase vector, the realization of this method consists of four steps as follows:

1. Generate and select N locations for fireworks randomly in the feasible space $[0 \ 2\pi]$.

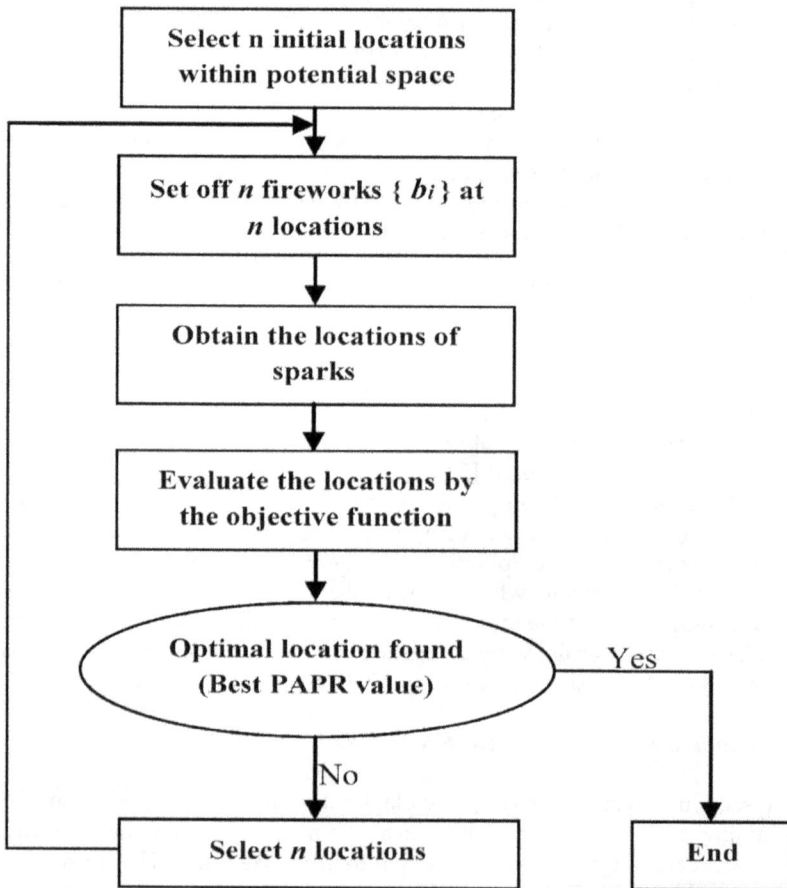

Figure 6.
Flow chart of fireworks algorithm.

2. Evaluate or calculate the fitness value of each firework according to the fitness function $(f_{obj}(\mathbf{b}_v))$. The number of sparks is calculated based on theory formula [21, 57] where the fireworks with better fitness values produce more sparks.

3. The position of sparks is controlled by The explosion amplitude which is determined by the fitness value of that firework [52, 57], each one represents a solution in the feasible space $[0 \ 2\pi]$. In general, the explosion amplitude for the firework with better fitness value is smaller and vice versa. Gaussian mutation is used to keep the diversity of the population in each iteration.

4. Calculate the best fitness value using objective function. If the terminal condition is met (number of iteration or best PAPR value), stop the algorithm. Otherwise, continue the iteration process. Based on selection strategy the best sparks are selected to form a new population.

In this section, many simulations have been performed based on IEEE 802.11a (Wireless LAN) to verify the performance of PTS-OFDM based on Fireworks algorithm. FWA is used to find the optimal combination of phase factors to reduce PAPR. The OFDM system was simulated with 64 subcarriers, in which 4 sub-blocks are employed, and the PTS, selected mapping (SLM) [58] and GA used W = 4 {1, −1, j, −j} phase weighting factors to optimize the PAPR of the modulated OFDM symbol, while others algorithms chose randomly the four phases within the interval W = $[0, 2\pi]$. For the FWA, the parameters were chosen as described in [52], the FWA worked quite well at the setting: $\hat{m} = 5$, $\hat{A} = 40$, N = 5, m = 50, a = 0.04 and b = 0.8 [57].

In **Figure 7**, we compare the performance of the FWA-based PTS with the most widely used algorithms for phase optimization such as SPSO, GA, and SA, in terms

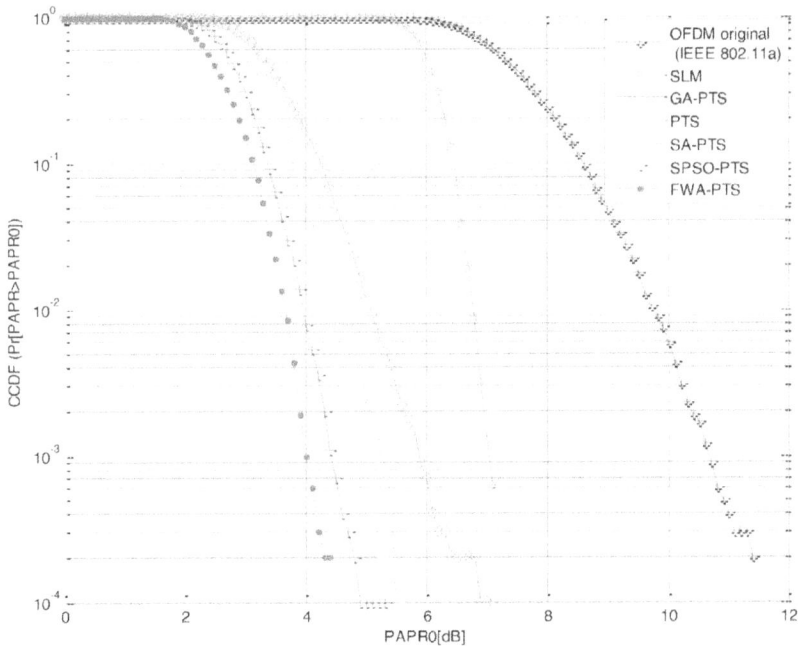

Figure 7.
PAPR reduction with PTS based different searching algorithms.

of CCDF (Complementary Cumulative Density Function). From **Figure** 7, it can be seen that the PTS-FWA scheme performs better than the other algorithms in terms of PAPR reduction. For example, at 10^{-3} of the CCDF, the PAPR is 4 dB, 4.421 dB, 4.948 dB, 5.226 dB, 5.879 dB, 7.034 dB, and 10.66 dB for the FWA, SPSO, SA, PTS, GA-PTS, SLM, and WLAN signals, respectively [57].

3.4.2 Improved versions of the fireworks algorithm

Fireworks Algorithm (FWA) is one of the best swarm intelligence algorithms, recently, many improvement versions of FWA have been proposed and developed based on several modifications. They were proposed to address some inherent limitations of the original algorithm.

Enhanced fireworks algorithm (EFWA) is an improved version of the recently developed Fireworks Algorithm (FWA) based on several modifications, it's proposed to tackle some limitations like the worse quality of the results when being applied on shifted functions or the high computational cost per iteration. In order to that, EFWA proposed five major improvements like a new minimal explosion amplitude check, a new operator for explosion, a new mapping strategy, a new operator for generating Gaussian sparks and for selecting the new population [53].

Dynamic fireworks algorithm (dynFWA) is an adaptive algorithm, it is an improved version of the recently developed EFWA based on an adaptive dynamic local search mechanism. DynFWA uses a dynamic explosion amplitude by increasing or decreasing the amplitude to speed up convergence when the fitness of the best firework could be improved (PAPR in our example) or to narrow the search area when the function could not be improved. In addition, DynFWA proposed the remove of Gaussian mutation operator to improve the computational efficiency [54].

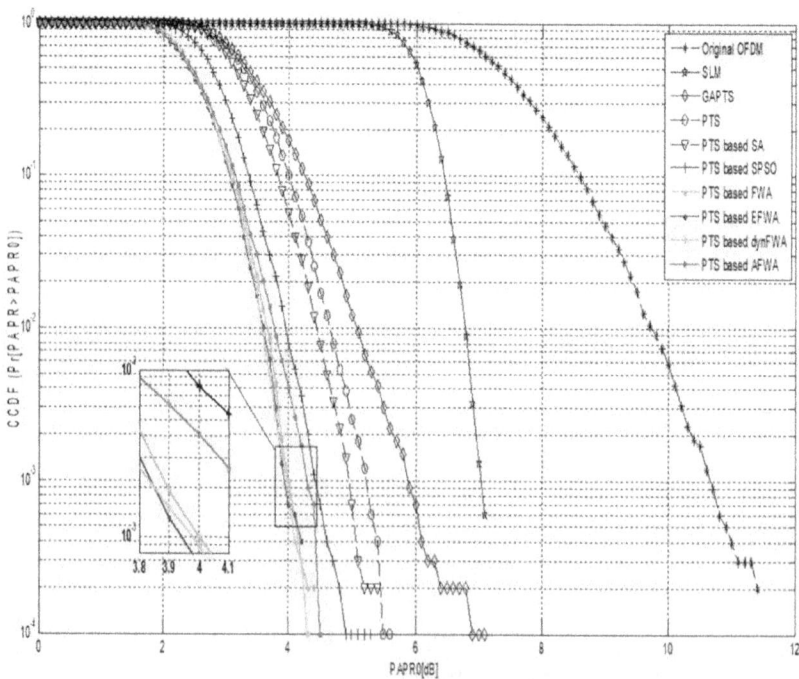

Figure 8.
PAPR reduction performance by the improved versions of the fireworks algorithm FWA.

Another new version called Adaptive fireworks algorithm (AFWA) is proposed to improve FWA and EFWA in term of the explosion amplitude which is a key factor influencing the performance of the algorithm. To improve the mechanism of calculating the amplitude, AFWA used the distance between the best firework and a certain selected individual as the explosion amplitude [55].

Figure 8 shows the simulation results for PAPR reduction of WLAN signals using the Fireworks algorithm based PTS technique and the recently improved versions of FWA (EFWA, DynFWA, and AFWA), compared with the different optimization approaches. From this figure, it is clear that FWA and all developed versions can effectively reduce PAPR in the WLAN-OFDM system. However, their PAPR reduction performance is different, in general, EFWA and dynFWA show a small improvement over the conventional FWA. For $CCDF = 10^{-3}$ we have 3.942 dB and 3.979 dB for EFWA and dynFWA, respectively while AFWA gives 4.283 dB and FWA 4 dB [57].

4. Computational complexity comparison

Beside the optimization accuracy, the convergence speed is an essential parameter for any optimization algorithm. To compare the convergence speed of SI algorithms, we performed some simulations shown in **Figure 9**, which represent the convergence curves of the FWA schemes in comparison with GA and SPSO.

The simulations are performed on a random OFDM symbol with 10 independent generation cycles and 3000 iterations. From these results, we can conclude that the four proposed FWA methods have a much faster convergence speed than SPSO and GA. **Table 2** shows that the fireworks algorithm and its improved versions can find optimal solutions in less than 500 function evaluations.

In terms of computational cost, **Figure 10** shows the time consumed by each algorithm to reduce PAPR. As an experimental platform we used some calculation software, run with a Win 7 operating system on an Intel (R) Core (TM) i5-2430M; 2.4 GHz; and 4 GB of RAM. As we see, the EFWA, dynFWA, and AFWA algorithms

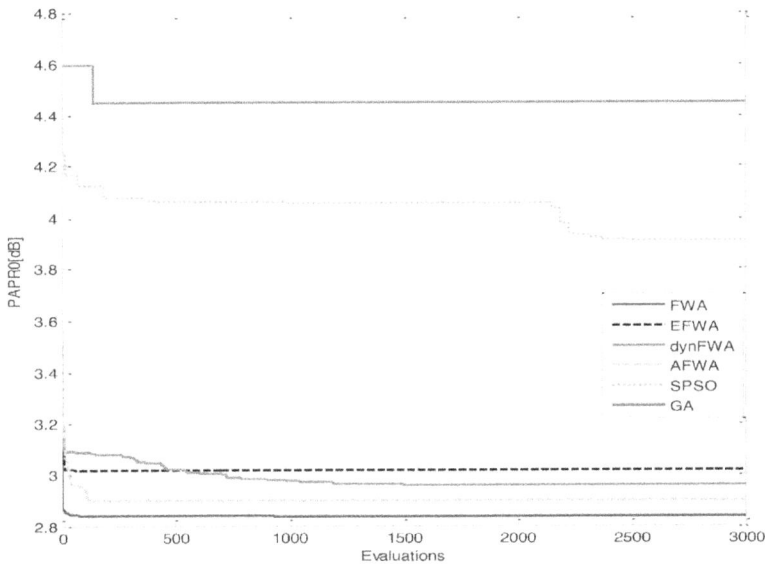

Figure 9.
Convergence curves of the SI algorithms for an OFDM symbol.

Methods	Function evaluations	PAPR [dB]
Original	—	10.66
SPSO	2000	4.055
GA	2000	4.447
FWA	500	2.839
EFWA	500	3.015
DynFWA	500	3.021
AFWA	500	2.9

Table 2.
Performance evaluation by the SI algorithms, on one symbol OFDM over 10 independent runs.

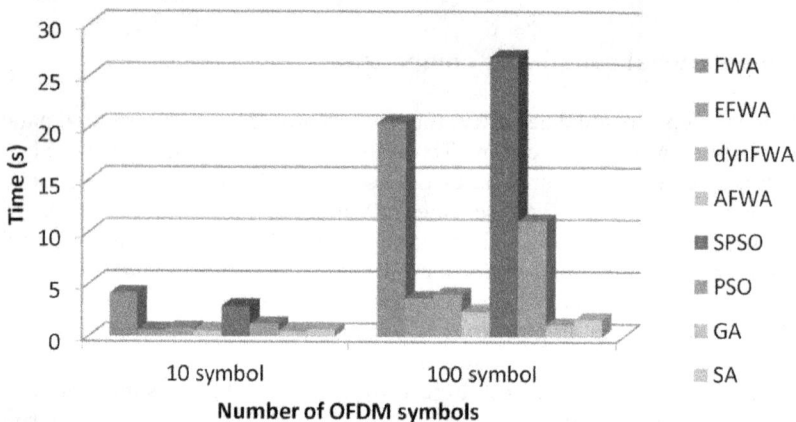

Figure 10.
Time consumed by each algorithm.

are close to each other in execution time, which is much shorter than the FWA and SPSO runtimes.

From these results, we can conclude that EFWA and AFWA have the best computational cost than FWA and dynFWA, while AFWA and dynFWA are very promising compared to other algorithms because of their efficiencies and simplicities.

5. Conclusion

In this chapter, we tried to present the performance of some optimization algorithms based PTS technique to reduce the PAPR of OFDM system with low computational complexity. First of all, OFDM, PAPR problem, and PTS scheme were presented to clarify the problem. Then a concise review on swarm intelligence domain was investigated. In others sections, a brief introduction to GA, PSO, SA and FWA is presented with primary focuses on the basic principal, algorithm study, problem solving, and some applications. Theoretical analysis is also described with completed reference citations of each algorithm. The SI algorithms were compared in terms of CCDF, and simulation results show that, FWA had better performance compared to GA, SA and SPSO. Some improved version of FWA, like EFWA, dynFWA, AFWA were also briefly described, and the comparison show that the new versions have a promising performance in both optimization accuracy of PAPR and convergence speed over conventional schemes and algorithms.

Conflict of interest

The authors do not have any conflicts of interest to declare.

Appendices and nomenclature

ACO	Ant Colony Optimization
AFWA	Adaptive fireworks algorithm
BER	Bit Error Rate
CCDF	Complementary Cumulative Distribution Function
DynFWA	Dynamic fireworks algorithm
EFWA	Enhanced fireworks algorithm
FWA	Fireworks algorithm
GA	Genetic Algorithm
HPA	High Power Amplifier
IFFT	Inverse Fast Fourier Transform
OFDM	Orthogonal Frequency Division Multiplexing
PAPR	Peak to Average Power Ratio
PSO	Particle Swarm Optimization
SI	Swarm Intelligence
PTS	Partial Transmit Sequence
SLM	Selective Mapping
$x[n]$	Discrete time signal
L	Oversampling factor
X_k	The input complex symbols
$E[.]$	The expected value
V	Number of sub blocks in PTS technique
b^v	Optimized phase vector in PTS technique
W	Phase Weighting Factors
N	Number of subcarriers
v_i	Velocity
f_{obj}	Objective function

Author details

Lahcen Amhaimar[1]*, Ali Elyaakoubi[2], Mohamed Bayjja[1], Kamal Attari[1] and Saida Ahyoud[3]

1 Modeling and Simulation of Intelligent Industrial Systems laboratory, Information technology and Artificial intelligence Team/ENSAD, Hassan II University of Casablanca, Morocco

2 LISAC laboratory, Faculty of science Dhar El Mahraz, Sidi Mohamed Ben Abdellah University-Fes, Morocco

3 Information Technology and Systems Modeling Team, Faculty of Sciences, Abdelmalek Essaadi University, Tetuan, Morocco

*Address all correspondence to: lahcen.amhaimar@univh2c.ma

IntechOpen

References

[1] S. Garnier, J. Gautrais, et G. Theraulaz, « The biological principles of swarm intelligence », *Swarm Intell.*, vol. 1, n° 1, p. 3-31, 2007.

[2] S. Das, A. Abraham, et A. Konar, « Swarm intelligence algorithms in bioinformatics », in *Computational Intelligence in Bioinformatics*, Springer, 2008, p. 113-147.

[3] A. E. Eiben et J. E. Smith, *Introduction to evolutionary computing*, vol. 53. Springer, 2003.

[4] Y. Tan et J. Wang, « Nonlinear blind source separation using higher order statistics and a genetic algorithm », *IEEE Trans. Evol. Comput.*, vol. 5, n° 6, p. 600-612, 2001.

[5] J. Zhang, L. Ni, C. Xie, Y. Tan, et Z. Tang, « Amt-pso: An adaptive magnification transformation based particle swarm optimizer », *IEICE Trans. Inf. Syst.*, vol. 94, n° 4, p. 786-797, 2011.

[6] P. J. Van Laarhoven et E. H. Aarts, « Simulated annealing », in *Simulated annealing: Theory and applications*, Springer, 1987, p. 7-15.

[7] M. T. Hagan, H. B. Demuth, et M. Beale, *Neural network design*. PWS Publishing Co., 1997.

[8] G. Ruan et Y. Tan, « A three-layer back-propagation neural network for spam detection using artificial immune concentration », *Soft Comput.*, vol. 14, n° 2, p. 139-150, 2010.

[9] Y. Tan et Z. Liu, « On matrix eigendecomposition by neural networks », *Neural Netw. World*, vol. 8, n° 3, p. 337-352, 1998.

[10] F. W. Glover et M. Laguna, *Tabu Search*. Springer US, 1997. doi: 10.1007/ 978-1-4615-6089-0.

[11] H.-O. Peitgen, H. Jürgens, et D. Saupe, « The Backbone of Fractals: Feedback and the Iterator », in *Chaos and Fractals*, Springer, 2004, p. 15-59.

[12] G. J. Klir et B. Yuan, « Fuzzy sets and fuzzy logic theory », *2nd Boston Kluwer Acad. Publ.*, 1995.

[13] J. Kennedy et R. Eberhart, « Particle swarm optimization », in *Proceedings of ICNN'95-international conference on neural networks*, 1995, vol. 4, p. 1942-1948.

[14] M. Dorigo, M. Birattari, et T. Stutzle, « Ant colony optimization », *IEEE Comput. Intell. Mag.*, vol. 1, n° 4, p. 28-39, 2006.

[15] D. Karaboga et B. Basturk, « A powerful and efficient algorithm for numerical function optimization: artificial bee colony (ABC) algorithm », *J. Glob. Optim.*, vol. 39, n° 3, p. 459-471, 2007.

[16] D. Karaboga et B. Basturk, « On the performance of artificial bee colony (ABC) algorithm », *Appl. Soft Comput.*, vol. 8, n° 1, p. 687-697, 2008.

[17] C. J. Bastos Filho, F. B. de Lima Neto, A. J. Lins, A. I. Nascimento, et M. P. Lima, « Fish school search », in *Nature-inspired algorithms for optimisation*, Springer, 2009, p. 261-277.

[18] C. R. Blomeke, S. J. Elliott, et T. M. Walter, « Bacterial survivability and transferability on biometric devices », in *Security Technology, 2007 41st Annual IEEE International Carnahan Conference on*, 2007, p. 80-84.

[19] X.-S. Yang, « Firefly algorithms for multimodal optimization », in *International symposium on stochastic algorithms*, 2009, p. 169-178.

[20] X.-S. Yang, « A new metaheuristic bat-inspired algorithm », in *Nature inspired cooperative strategies for optimization (NICSO 2010)*, Springer, 2010, p. 65-74.

[21] Y. Tan et Y. Zhu, « Fireworks algorithm for optimization », *Adv. Swarm Intell.*, p. 355-364, 2010.

[22] Y. Shi, « Brain Storm Optimization Algorithm », in *Advances in Swarm Intelligence*, Berlin, Heidelberg, 2011, p. 303-309. doi: 10.1007/978-3-642-21515-5_36.

[23] M.-H. Tayarani-N et M. R. Akbarzadeh-T, « Magnetic optimization algorithms a new synthesis », in *2008 IEEE Congress on Evolutionary Computation (IEEE World Congress on Computational Intelligence)*, 2008, p. 2659-2664.

[24] H. Shah-Hosseini, « The intelligent water drops algorithm: a nature-inspired swarm-based optimization algorithm », *Int. J. Bio-Inspired Comput.*, vol. 1, n° 1-2, p. 71-79, 2009.

[25] D. E. Goldberg, « Genetic algorithms in search, optimization, and machine learning, 1989 », *Read. Addison-Wesley*, 1989.

[26] C. Ergun et K. Hacioglu, « Multiuser detection using a genetic algorithm in CDMA communications systems », *IEEE Trans. Commun.*, vol. 48, n° 8, p. 1374-1383, 2000.

[27] S. P. DelMarco, « A Constrained Optimization Approach to Compander Design for OFDM PAPR Reduction », *IEEE Trans. Broadcast.*, vol. 64, n° 2, p. 307-318, 2018.

[28] S. H. Han et J. H. Lee, « An overview of peak-to-average power ratio reduction techniques for multicarrier transmission », *IEEE Wirel. Commun.*, vol. 12, n° 2, p. 56-65, 2005.

[29] L. Amhaimar, S. Ahyoud, et A. Asselman, « An efficient combined scheme of proposed PAPR reduction approach and digital predistortion in MIMO-OFDM systems », *Int. J. Commun. Antenna Propag.*, vol. 7, n° 5, 2017.

[30] H.-S. Joo, K.-H. Kim, J.-S. No, et D.-J. Shin, « New PTS schemes for PAPR reduction of OFDM signals without side information », *IEEE Trans Broadcast*, vol. 63, n° 3, p. 562-570, 2017.

[31] L. J. Cimini et N. R. Sollenberger, « Peak-to-average power ratio reduction of an OFDM signal using partial transmit sequences », *IEEE Commun. Lett.*, vol. 4, n° 3, p. 86-88, 2000.

[32] A. Joshi et D. S. Saini, « Peak-to-Average Power Ratio Reduction of OFDM signals Using Improved PTS Scheme with Low Computational Complexity », *WSEAS T Commun*, vol. 12, p. 630-640, 2013.

[33] R. W. Bauml, R. F. Fischer, et J. B. Huber, « Reducing the peak-to-average power ratio of multicarrier modulation by selected mapping », *Electron. Lett.*, vol. 32, n° 22, p. 2056-2057, 1996.

[34] S. H. Muller et J. B. Huber, « OFDM with reduced peak-to-average power ratio by optimum combination of partial transmit sequences », *Electron. Lett.*, vol. 33, n° 5, p. 368-369, 1997.

[35] L. Amhaimar, S. Ahyoud, et A. Asselman, « Peak-to-average power ratio reduction in MIMO-OFDM systems », *Int. J. Microw. Opt. Technol.*, vol. 12, n° 1, p. 9-16, 2017.

[36] J. Robinson et Y. Rahmat-Samii, « Particle swarm optimization in electromagnetics », *IEEE Trans. Antennas Propag.*, vol. 52, n° 2, p. 397-407, 2004.

[37] T. Jiang, W. Xiang, P. C. Richardson, J. Guo, et G. Zhu, « PAPR

reduction of OFDM signals using partial transmit sequences with low computational complexity », *IEEE Trans. Broadcast.*, vol. 53, n° 3, p. 719-724, 2007.

[38] M. Lixia, M. Murroni, et V. Popescu, « PAPR reduction in multicarrier modulations using Genetic Algorithms », in *Optimization of Electrical and Electronic Equipment (OPTIM), 2010 12th International Conference on*, 2010, p. 938-942.

[39] H. Liang, Y.-R. Chen, Y.-F. Huang, et C.-H. Cheng, « A modified genetic algorithm PTS technique for PAPR reduction in OFDM systems », in *2009 15th Asia-Pacific Conference on Communications*, 2009, p. 182-185.

[40] A. Ratnaweera, S. K. Halgamuge, et H. C. Watson, « Self-organizing hierarchical particle swarm optimizer with time-varying acceleration coefficients », *IEEE Trans. Evol. Comput.*, vol. 8, n° 3, p. 240-255, 2004.

[41] M. Clerc et J. Kennedy, « The particle swarm-explosion, stability, and convergence in a multidimensional complex space », *IEEE Trans. Evol. Comput.*, vol. 6, n° 1, p. 58-73, 2002.

[42] W.-C. Liu, « Design of a multiband CPW-fed monopole antenna using a particle swarm optimization approach », *IEEE Trans. Antennas Propag.*, vol. 53, n° 10, p. 3273-3279, 2005.

[43] K. Attari, L. Amhaimar, A. E. Yaakoubi, et A. Asselman, « InP/Si High Efficiency Heterojunction-Junction Solar Cell Design Using PSO and the GA Algorithms », *Int. Rev. Electr. Eng. IREE*, vol. 15, n° 3, Art. n° 3, juin 2020, doi: 10.15866/iree.v15i3.17155.

[44] A. Khare et S. Rangnekar, « A review of particle swarm optimization and its applications in solar photovoltaic system », *Appl. Soft Comput.*, vol. 13, n° 5, p. 2997-3006, 2013.

[45] D. Bratton et J. Kennedy, « Defining a standard for particle swarm optimization », in *Swarm Intelligence Symposium, 2007. SIS 2007. IEEE*, 2007, p. 120-127.

[46] Y. Tan et Z. M. Xiao, « Clonal particle swarm optimization and its applications », in *Evolutionary Computation, 2007. CEC 2007. IEEE Congress on*, 2007, p. 2303-2309.

[47] D. Abramson, « A very high speed architecture for simulated annealing », *Computer*, vol. 25, n° 5, p. 27-36, 1992.

[48] P. Banerjee, M. H. Jones, et J. S. Sargent, « Parallel simulated annealing algorithms for cell placement on hypercube multiprocessors », *IEEE Comput. Archit. Lett.*, vol. 1, n° 01, p. 91-106, 1990.

[49] A. Dekkers et E. Aarts, « Global optimization and simulated annealing », *Math. Program.*, vol. 50, n° 1, p. 367-393, 1991.

[50] G. Dueck et T. Scheuer, « Threshold accepting: A general purpose optimization algorithm appearing superior to simulated annealing », *J. Comput. Phys.*, vol. 90, n° 1, p. 161-175, 1990.

[51] T. M. Nabhan et A. Y. Zomaya, « A parallel simulated annealing algorithm with low communication overhead », *IEEE Trans. Parallel Distrib. Syst.*, vol. 6, n° 12, p. 1226-1233, 1995.

[52] Y. Tan, *Firework Algorithm: A Novel Swarm Intelligence Optimization Method*. Springer, Berlin, Heidelberg, Germany, 2015.

[53] S. Zheng, A. Janecek, et Y. Tan, « Enhanced fireworks algorithm », in *Evolutionary Computation (CEC), 2013 IEEE Congress on*, 2013, p. 2069-2077.

[54] S. Zheng, A. Janecek, J. Li, et Y. Tan, « Dynamic search in fireworks

algorithm », in *Evolutionary Computation (CEC), 2014 IEEE Congress on*, 2014, p. 3222-3229.

[55] J. Li, S. Zheng, et Y. Tan, « Adaptive fireworks algorithm », in *Evolutionary Computation (CEC), 2014 IEEE Congress on*, 2014, p. 3214-3221.

[56] L. Amhaimar, A. El Yaakoubi, M. El Halaoui, mohamed Bayjja, M. E. H. Hajri, et S. Ahyoud, « A New Approach of PAPR Reduction with Low Computational Complexity in MIMO-OFDM Systems Based Smart Optimization Algorithm », *Int. J. Microw. Opt. Technol.*, vol. 14, n° 2, p. 116-123, mars 2019.

[57] L. Amhaimar, S. Ahyoud, A. Elyaakoubi, A. Kaabal, K. Attari, et A. Asselman, « PAPR Reduction Using Fireworks Search Optimization Algorithm in MIMO-OFDM Systems », *J. Electr. Comput. Eng.*, vol. 2018, p. e3075890, sept. 2018, doi: 10.1155/2018/3075890.

[58] X. Cheng, D. Liu, S. Feng, H. Fang, et D. Liu, « An artificial bee colony-based SLM scheme for PAPR reduction in OFDM systems », in *2017 2nd IEEE International Conference on Computational Intelligence and Applications (ICCIA)*, 2017, p. 449-453.